GENETICS, EVOLUTION,

RACE,

RADIATION BIOLOGY

ESSAYS IN
SOCIAL BIOLOGY
Volume II

PRENTICE-HALL INTERNATIONAL, INC., *London*
PRENTICE-HALL OF AUSTRALIA, PTY. LTD., *Sydney*
PRENTICE-HALL OF CANADA, LTD., *Toronto*
PRENTICE-HALL OF INDIA PRIVATE LIMITED, *New Delhi*
PRENTICE-HALL OF JAPAN, INC., *Tokyo*

GENETICS, EVOLUTION,

RACE,

RADIATION BIOLOGY

ESSAYS IN
SOCIAL BIOLOGY
Volume II

Bruce Wallace

Division of Biological Sciences
Cornell University

PRENTICE-HALL, INC.
Englewood Cliffs, New Jersey

10   9   8   7   6   5   4   3   2   1

A/301. 3ʹ/

ISBN :   0:13-351155-3 (p)   0-13-351163-4 (c)
Library of Congress Catalogue Card Number 79-167789
Printed in the United States of America

To   **MCW**

        DBW

        RSW

Undoubtedly the most momentous piece of progress of 1969 occurred in Dijon, France, on August 14 when Neil Rappaport of Festoona, N.J., lost forever the blueprints for a machine he had just invented which would have made it possible to create a louder noise than had ever been heard before.

Acknowledgments

The author wishes to acknowledge that these essays owe their existence to a criticism made by Professor Jacques Barzun of college-level "science survey" courses. Whether I have succeeded in meeting this criticism is, of course, another matter.

A large number of colleagues at Cornell University have helped me in one way or another during the preparation of this book. Dr. Richard B. Root outlined the fascinating aspects of ecology which relate to the selection from *Iberia*. Similarly, Dr. Thomas Eisner helped me to identify many points of interest concerning communication. Dr. Richard D. O'Brien aided me in understanding the details of the transmission of nerve impulses and the modification of this transmission by chemical means. In many conversations, Dr. Gerald R. Fink explained matters related to diseases, their transmission, and the action of antibiotics. Indeed, virtually all persons who have taken part in the *Biology and Society* lectures at Cornell University have aided me immensely.

No less than my colleagues on the faculty, I must also thank the students of *Biology and Society* who undertook to read and criticize the essays during their preparation. It is a pleasure to acknowledge with special thanks the efforts of Cynthia Ravitski, Terri Schwartz, and Mark Zabek. Penny Farrow and Florence Robinson also had numerous encouraging comments and useful criticisms during this same period.

Within my own family, each member read (and seemingly enjoyed) one or more essays. Considered with my family are a host of college students who were unexpectedly pressed into reading these essays while visiting our home — both in Ithaca and abroad during the summer of 1970. Every reaction whether heeded or not has helped in the preparation of this book. And each and every one has been appreciated.

Finally, I must thank Drs. Franklin A. Long and Raymond Bowers of the Program for Science, Technology, and Society at Cornell University who arranged for a release of two months from my regular duties so that I could complete these essays. I wish to thank Dr. Harry T. Stinson for his efforts in consummating these arrangements. Financial support during these months was provided through a grant from the Alfred P. Sloan Foundation. The Program for Science, Technology, and Society has been instrumental, as well, in supporting the *Biology and Society* lectures at Cornell, a series of talks on topics similar to those discussed in my own essays.

Contents

Preface

This book is like no other biology text I know. I say this immediately to warn the reader, both my professional colleagues and the student. These essays have been written to be enjoyed. I assume that the reader has had a modern high school biology course, that he enjoys reading, and that his instructor will fill in or enlarge upon background factual material wherever necessary. Proceeding from these assumptions, I have attempted to explore contemporary biology by means of single-topic essays — touching upon a variety of questions in as informal and nontechnical a manner as possible.

From classroom experience I know that the student's initial reaction to this text is one of discomfort. Where, he asks, is the glossary of terms from which routine quizzes are prepared? What can I memorize? What am I expected to learn? What will the professor ask on his examinations? These are the conditioned reflexes of today's student — reflexes honed to a fine edge during many years of secondary and higher education. Because of the exaggerated emphasis that is placed by schools on measurable performance, much of the student's commitment to any knowledge that he might have gained in class is shed as he hands in his final exam; at that moment he has passed or failed in his effort to please his professor. Next comes the pleasure of forgetting most of the unpleasant ordeal.

Today's serious problems are not in the classroom; they are outside. These problems demand an aroused public equipped with all the information and intellectual skill that individuals can muster. We can no longer afford to shuck our knowledge at the classroom door as we depart. My plea to both the student and the instructor, therefore, is to abandon the weekly vocabulary drill and the multiple-choice examination; instead, read the essays and their accompanying selections, enjoy them, react to them, and through prolonged discussion criticize and rethink them. If enjoyment and critical discussion cause a student to retain some of what he has learned, fine. We must by all means encourage him to enjoy biology, for it is only because he may later make use of what he has retained that we teach biology to the nonbiologist.

Before World War II, bird watching and nature study made an adequate biological background for someone whose daily efforts were spent in court, in a business office, in a factory, or teaching one of the humanities in high school or college. Biology of this sort offered a pleasant diversion to the business of earning a living. Simple biology, however, is no longer sufficient. Today, many of man's problems have a biological basis — numbers of persons, the production of food to feed the hungry, man's extermination of other living things, and the wholesale destruction of the environment by man's industry. None of us can afford any longer to be a mere spectator; where a few fight for immediate profit, the many must fight for a lasting place to live. Many problems – racial, genetic, eugenic – lurking on the horizon are also biological ones. It is ridiculous to expect intelligent action to come from biologically naive persons who, with fatal optimism, seek at most ad hoc solutions to grave problems. The decisions affecting irreversible governmental and industrial actions are largely scientific in nature; we can no longer be satisfied with science survey courses that merely train dillettante bird watchers and butterfly collectors. These courses must teach students to grapple with the serious problems of today – and tomorrow.

What precisely is the aim of a required course in elementary biology at the college level? What is it that the student should learn? Opinions differ. I believe that today's college student most likely has had a modern course in high school biology comparable to those prepared by the Biological Sciences Curriculum Study. I feel that he should understand the workings of his body and mind ("know thyself" is still an apt phrase) and how both continue to work under a barrage of external and internal challenges. He should see how mankind fits into the skein of life on earth and appreciate the constraints placed upon man if both he and the skein are to continue their existence. We are, as the astronauts said, confined to a raft that is a small blue gem floating in space; our problem is to keep the raft gemlike and blue. Finally, the student should develop a sense of understanding and a method of intellectual attack appropriate for the problems that confront (haunt?) us in our daily relationship with government, business, community, and neighborhood affairs. The continual reliance upon expedient solutions to grave problems resembles the instant-by-instant reflexive responses of a frightened hare attempting to elude a dog; it does not become the intellectual powers of man. A college-level biology course for nonmajors should help the individual visualize and evaluate probable outcomes of certain courses of action. It should teach him to cope with and refute the oldest and most seductive of all ad hoc arguments: "It smells like money."

How does one obtain these goals? In truth, no one knows. To ask each student to make personal observations on all matters of import is patently foolish; in our complex society such observations must be made vicariously. Consequently, in the sections that follow I have drawn upon the observations and experiences made and described not by scientists but by essayists, novelists, and historians. I have turned to these persons because they are not only professional observers but also are competent writers. In doing this I may have risked scientific precision but I have gained tremendously in other respects. A clinical account, no matter how accurate, cannot match Voltaire when he writes "a beggar all covered with sores, dead-eyed, the end of his nose eaten away, his mouth deformed, his teeth black, tormented with a violent cough . . . ."

Because students are initially discomforted by my approach, I want to explain further my use of literary selections in a science text. In my view they substitute for personal observations but not for the precise and structured observations of a field trip. They more nearly resemble series of strolls, some brief and others perhaps overly long. Each establishes a mood and offers an excuse to embark upon a series of discussions; the essays serve this latter purpose. No attempt has been made to write an essay about every possible topic posed by its accompanying selection nor have the essays been restricted to topics included within the selection. If the selections can be compared with strolls, the essays can be compared with the conversational ramblings of a talkative companion – one inspired by events he encounters but whose conversation is neither restricted to nor necessarily evoked by the passing scene.

The reader will find in these three volumes selections from a number of sources as well as a variety of single-topic essays. Their arrangement, of necessity, is not rigorously sequential. Sometimes in earlier essays I have omitted details needed to understand later ones. Sometimes I have repeated information. In any case, the effect is to make the essays just the more chatty and repetitious. When the reader has finished the last essay, he will be totally uninformed of such things as the difference between *polyploidy, polyteny,* and *polysomy,* but why shouldn't he be? There is a remote chance that he will have enjoyed himself at some point. If his enjoyment causes him to retain a knack for wrestling with today's problems, and if he uses this knack by speaking up publicly on controversial environmental and population issues, his enjoyment could prove to be the lasting contribution of the book.

Bruce Wallace

A Word to the Teacher

The use of this text will call for techniques and attitudes unlike those found in most biology classes. Student interest in the types of topics discussed in these essays is nearly overwhelming. Lecturers in the initial *Biology and Society* series at Cornell University (Biology 201-202, 2 points per semester) found themselves addressing sustained weekly audiences of 1,000 (students and townspeople) rather than the 25 or 30 that most had expected. The problems that arise from an enthusiastic response of this kind are of three sorts: (1) the handling of large numbers of students in an intimate way; (2) the choice of factual material to be stressed in the course; (3) the basis of grading students who enroll for credit.

How can large numbers of students be handled where the burden of a social biology course falls on one staff member? I suggest that this man (presumably a biologist) be conversant with the material covered by any one of a number of excellent college-level biology textbooks and, further, with the additional material to be found in some 130 *Scientific American* offprints. Each lecture should be restricted to no more than 20 minutes; it should present only the facts that are essential for understanding the assigned essay(s). The remaining 30 minutes should be reserved for student discussion and questions from the floor. The role of the lecturer during this period is to encourage student participation and to supply factual information when necessary in order to keep the discussion accurate. Where information is lacking or when someone questions the accuracy of an essay, a team of two or three students should be told to recheck the facts and return in a week's time with a five- or ten-minute report for the class.

Facts and the collection of factual material should be recurrent themes throughout all discussions. The types of facts, however, that are important in a type of course like *Biology and Society* need not be those that are considered important in most beginning biology courses. Vocabulary is trivial, except to the biology major whose livelihood will depend upon knowing it. Many of the details of ordinary biology courses are also trivial — especially those details that are rapidly forgotten by most students and, consequently, are so useful as test items. The facts that I feel are important are the obvious ones that tend to be neglected. These are the facts that at times are embarrassing to teach because the students react by saying, "Of course." They must be repeated again and again nevertheless. If simple facts were understood by everyone, permanent equipment would not be designed to wear out nor would disposable articles be made so often of indestructible materials.

My personal feelings on the matter of grades and grading are rather strong ones. The student who attends a social biology course and later, because of it, takes an active and intelligent part in community affairs attests to the success of the course and its teacher; the student who memorizes facts, talks in class, but refuses to become involved in society's affairs represents a failure. I have argued that grades for Cornell's *Biology and Society* should be limited to S/U (satisfactory-unsatisfactory) *only* and that those who take the course should receive credit only for use in accumulating general credits. This argument is based in part on my feeling that the choice of lecture topics and the overall format of the course can most likely remain flexible and stimulating only if traditional administrative groups and committees are kept uninvolved; once a standing committee feels that the course content falls within its domain, freedom to arrange novel discussions or sequences of off-beat lecture topics may disappear. In addition, however, I firmly believe that no appropriate grade in social biology can be given until at least two decades after the student's graduation.

SECTION ONE

Genetics

Introduction

Because molecular biology has provided the spectacular scientific advances of the second half of the 20th century, most persons are familiar with the terms deoxyribonucleic acid (DNA), ribonucleic acid (RNA), the Watson-Crick model of DNA structure, and the genetic code. The selections that introduce this section have been written by one who from his student days has been intimately associated with the development of molecular genetics: James D. Watson.

The selections from Watson's writings have been chosen not for their description of the growth of the science of molecular genetics but for what they reveal of the development of Watson and of his relation to those about him. Here we find in the company of eminent theoretical and experimental biologists the brilliant young graduate student struggling to make sense of observations, striving to see the invisible connections between facts that must by their nature be interconnected, and dreaming of devising *the* general scheme within which the multitude of experimental details would make sense – like the bits and pieces of a completed jig-saw puzzle. Struggling and striving to be – and dreaming of being – first.

Watson was successful. A regime of hard thought and hard play (the latter may be by far the more obvious to an outsider and is likely to leave the impression that scientists are a rather juvenile lot) took him from Chicago to Cold Spring Harbor to Pasadena to Copenhagen to Naples to Cambridge and, finally, to Stockholm for the Nobel Prize. And, as he describes his voyage, we learn to our delight that at one moment, Watson used his sister as a lure while bettering his own position in the great DNA-race.

●　●　●

The essays that complete the section deal for the most part with general aspects of genetics, especially the relation between what we now know and what had been surmised by H. J. Muller and earlier geneticists on purely theoretical grounds. The demands made of genetic material if it were to fulfill known hereditary functions and the means by which DNA meets these demands make the combined studies of Watson, Crick, and their many colleagues and predecessors an eminently satisfying fragment of the large mural of science.

In some respects the following essays will be disconcerting; many readers will assume that the logic of genetics has led me to write difficult and abstract essays in this section instead of the low-key, largely anecdotal ones of earlier sections. These persons should not be misled by the change in style. In one of the earlier essays, I took the reader by the hand and led him through an exponential sequence:

1 - 2 - 4 - 8 - 16 - 32 - 64 - 128 - 256 - 512 - 1024 -

I did this not because the reader is unable to recite the sequence by himself but because when he does, he recites it as a meaningless exercise divorced from life's experiences. In response to a recent questionnaire, 80 per cent of the people in one university community said that they favor a stable population size (zero population growth), but in response to still another question, 80 per cent of the same persons said they favored families with three children. As long as highly educated persons hold such self-contradicting views there is a need for someone to take the lead in reciting

- 2048 - 4096 - 8192 - 16384 - 32768 - 65536 - 131072 -

Neglect of the obvious, not ignorance of the difficult, will prove to be man's undoing.

Molecular genetics, DNA, self-replication, gene control, and other aspects of modern biology are topics discussed in daily newspapers and weekly news magazines. Practicing scientists frequently learn of their colleagues' activities from the paper at breakfast. High-school texts have been constructed around the central dogma of molecular genetics. Biology teachers are at any moment prepared to expound on the intricacies of Mendelian ratios. Consequently, I have felt no compulsion to trudge through elementary genetic facts in this section; rather, I have written on and about matters in a manner that is intended to emphasize the past of genetics as well as its present, to supplement the instructor rather than replace him.

Growing Up

in the Phage Group*

James D. Watson

As an undergraduate at Chicago, I had already decided to go into genetics even though my formal training in it was negligible, with most of my course work reflecting a boyhood interest in natural history. Population genetics at first intrigued me, but from the moment I read Schrodinger's "What is Life" I became polarized toward finding out the secret of the gene. My obvious choice for graduate school was Caltech, since I was told its Biology Division was loaded with good geneticists. They, however, did not want me, nor did Harvard, to which I had applied without considering what I might find. Harvard's disinterest in me was particularly fortunate, for if I had gone there I would have found no one excited by the gene and so might have been tempted to go back into natural history. Fortunately my advisor at Chicago, the human geneticist Strandskov, also had me apply to the Indiana University in Bloomington, emphasizing that H. J. Muller was there as well as several very good younger geneticists (Sonneborn and Luria). To my relief, Indiana took a chance with me, offering a $900 fellowship for the coming 1947-48 academic year. Characteristically, Fernandus Payne, then dean of its graduate school, wanted to make sure that I knew what I was getting into. He appended a postscript to the fellowship offer saying that if I wanted to continue my interest in birds I should go elsewhere.

During my first days at Indiana, it seemed natural that I should work with Muller but I soon saw that Drosophila's better days were over and that many of the best younger geneticists, among them Sonneborn and Luria, worked with micro-organisms. The choice among the various research groups was not obvious at first, since the graduate student gossip reflected unqualified praise, if not worship, of Sonneborn. In contrast, many students were afraid of Luria who had the reputation of being arrogant toward people who were wrong. Almost from Luria's first lecture, however, I found myself much more interested in his phages

*From John Cairns, Gunther S. Stent, and James D. Watson, Phage and the Origins of Molecular Biology. Cold Spring Harbor Laboratory, Cold Spring Harbor, New York, 1966.

than in the Paramecia of Sonneborn. Also, as the fall term wore on I saw no evidence of the rumoured inconsiderateness toward dimwits. Thus with no real reservations (except for occasional fear that I was not bright enough to move in his circle) I asked Luria whether I could do research under his direction in the spring term. He promptly said yes and gave me the task of looking to see whether phages inactivated by X rays gave any multiplicity reactivation.

The only other scientist in Luria's lab then was Renato Dulbecco, who six months previously had come from Italy to join in the experiments on the multiplicity reactivation of UV-killed [UV=ultraviolet radiation] phages. I was given a desk next to Dulbecco's and, when he was not doing experiments, often worked on his lab bench. Most of Luria's and Dulbecco's conversation was in Italian, and I might have felt somewhat isolated had it not been for the fact that my first experiments gave a slightly positive result. Usually, Luria never let even a few hours pass between the counting of my plaques and his knowing the answer. Also, Dulbecco's family had not yet come from Italy, and we would occasionally eat together at the Indiana Union. During one Sunday lunch, I remember asking him whether Luria's figure of 25 T2 genes should not tell us the approximate size of the gene since the molecular weight of T2 could be guessed from electron micrographs. Dulbecco, however, did not seem interested, perhaps because he already suspected that multiplicity reactivation of UV-killed phage was more messy than Luria's pretty subunit theory proposed. Then there was also the fact that despite Avery, McCarthy, and MacCleod, we were not at all sure that only the phage DNA carried genetic specificity.

Some weeks later in Luria's flat, I first saw Max Delbrück, who had briefly stopped over in Bloomington for a day. His visit excited me, for the prominent role of his ideas in "What is Life" made him a legendary figure in my mind. My decision to work under Luria had, in fact, been made so quickly because I knew that he and Delbrück had done phage experiments together and were close friends. Almost from Delbrück's first sentence, I knew I was not going to be disappointed. He did not beat around the bush and the intent of his words was always clear. But even more important to me was his youthful appearance and spirit. This surprised me, for without thinking I assumed that a German with his reputation must already be balding and overweight.

Then, as on many subsequent occasions, Delbrück talked about Bohr and his belief that a complementarity principle, perhaps like that needed for understanding quantum mechanics, would be the key to the real understanding of biology. Luria's views were less firm, but there was no doubt that on most days he too felt that the gene would not be simple and that high powered brains, like Delbrück's or that of the even more legendary Szilard, might be needed to formulate the new laws of physics (chemistry?) upon which the self replication of the gene was based. So, sometimes I worried that my inability to think mathematically might mean I could never do anything important. But in the presence of Delbrück I hoped I might someday participate just a little in some great revelation.

I looked forward greatly to the forthcoming summer (1948) when Dulbecco and I would go with the Lurias to Cold Spring Harbor. Delbrück and his wife Manny were coming for the second half while, before they arrived, there was to be the phage course given by Mark Adams. No great conceptual advances however, came out that summer. Nonetheless, morale was high even though Luria and Dulbecco sometimes worried whether they had multiplicity reactivation all wrong. Delbrück remained confident, however, that multiplicity reactivation was the key breakthrough which soon should tell us what was what. His attention, however, was then often directed toward convincing us that an argument of alternate steady states would explain Sonneborn's data on antigenic transformations in Paramecia. The idea of cytoplasmic hereditary determinants did not appeal at all to Delbrück and he hoped we would all join together to try to bury as many of them as possible.

As the summer passed on I liked Cold Spring Harbor more and more, both for its intrinsic beauty and for the honest ways in which good and bad science got sorted out. On Thursday evenings general lectures were given in Blackford Hall by the summer visitors and generally everyone went, except for Luria who boycotted talks on extra-sensory perception by Richard Roberts and on the correlation of human body shapes with disease and personality by W. Sheldon. On those evenings, as on all others, Ernst Caspari opened and closed the talks, and we marveled at his ability to thank the speakers for their "most interesting presentations."

Most evenings we would stand in front of Blackford Hall or Hooper House hoping for some excitement, sometimes joking whether we would see Demerec going into an unused room to turn off an unnecessary light. Many times, when it became obvious that nothing unusual would happen, we would go into the village to drink beer at Neptune's Cave. On other evenings, we played baseball next to Barbara McClintock's cornfield, into which the ball all too often went.

There was also the fair possibility that we could catch Seymour Cohen and Luria each informing the other that his experiments were not only over-interpreted but off the mainstream to genuine progress. Though Cohen was spending the summer doing experiments with Doermann, a sharp gap existed between Cohen and the phage group led by Luria and Delbrück. Cohen wanted biochemistry to explain genes, while Luria and Delbrück opted for a combination of genetics and physics.

Cohen was not, however, the only biochemist about. David Shemin spent most of the summer living in Williams House while Leonor Michealis stayed for several weeks, despite his wife's complaints about the run-down condition of their apartment and of Demerec's failure to replace a toilet seat containing a large crack. When August began the Lurias went home to Bloomington because Zella Luria would soon have a child. This left Dulbecco and me even more free to swim at the sand spit or to canoe out into the harbor often in search of clams or mussels.

By the time we were back in Bloomington, all of us were again ready for serious experiments. Soon Dulbecco found photoreactivation of UV inactivated phage, thereby explaining why the plaque counts in multiplicity reactivation experiments were often annoyingly inconsistent. This discovery did not seem a pure blessing, however, for it immediately threw into doubt all previous quantitative interpretation of multiplicity reactivation. Thus much of the work of the previous eighteen months had to be repeated, both in the light and dark. When this was finally done it became clear that multiplicity reactivation was, by itself, not going to yield simple answers about the genetic organization of phage. As a corollary, my study of X-ray inactivated phage also was much less likely to yield anything very valuable. By then, however, I had begun to study the indirect as well as the direct effect of X rays, and the complexity of the inactivation curves initially kept me from worrying whether they would be very significant.

That fall I had my first extended view of Szilard, when Luria, Dulbecco and I drove up to Chicago to see him and Novick. There I first realized that most conversations with Szilard occurred during meals, which seemingly consumed half of his time awake. Briefly I tried to tell him what I was up to, but soon I was crushed by his remark that I did not know how to speak clearly. Even more to the point was that Szilard did not like to learn new facts unless they were important or might lead to something important. Szilard and Novick later came to Bloomington for a small phage meeting in the spring of 1949. Hershey, Doermann, Weigle, Putnam, Kozoff and Delbrück were also there. Doermann had just done his genetic crosses using premature lysates and guessed that the percentage of recombinants was approximately constant throughout the latent period. Stent and Wollman described how they thought T4 interacted with tryptophan. To me the most memorable aspect of the meeting, however, was the unplanned comic performances of Szilard and Novick. Neither understood the other's description of their phenotypic mixing experiments and they were constantly interrupting each other, hoping to make the matter clear to everyone else. A day later, Delbrück, Luria, Delbecco and I drove to Oak Ridge for its spring meeting where Delbrück coined the phrase "The Principle of Limited Sloppiness" in explaining how Kelner and Dulbecco came upon photo-reactivation.

The following summer Manny Delbrück was expecting a child and most of the phage group congregated in Pasadena instead of Cold Spring Harbor. Several times each week, there occurred seminars dominated by Delbrück's insistence that the results logically fit into some form of pretty hypothesis. There were also innumerable camping trips occupying two to four days, long weekends often led by Carleton Gajdusek whose need for only two or three hours sleep a night allowed him to spend five or six days each week in the wilderness while maintaining the pretense that he was interested in the world between John Kirkwood and Delbrück. Because Gunther Stent shared a house in the San Gabriel foothills with Jack Dunitz, Pauling's postdoctoral student, there were frequent social contacts with the younger students who worked for Linus Pauling but, on the whole, I never got the impression that the phage group thought that Pauling's world and theirs would soon have anything in common. Occasionally, I would see Pauling drive up in his Riley and I felt very good when once he spontaneously smiled at me in the Faculty Club.

Most of the scientific arguments that summer were kinetic either about how tryptophan affected T4 adsorption or attempting to make sense of photoreactivation. Sometimes the genetic results of Hershey came into the picture but only Doermann seemed tempted to do more along that line. My experiments on X-ray phage had progressed to the point where I knew I had a thesis, and so in Pasadena I played about a little with formaldehyde inactivated phages. Delbrück, like everyone else, was only mildly interested in my results but told me that I was lucky that I had not found anything as exciting as Dulbecco had, thereby being trapped into a rat race where people wanted you to solve everything immediately. If that had happened, he felt I would lose in the long term by not having the time either to think or to learn what other people were doing. I of course wanted something important to emerge from the masses of survival curves that filled several thick loose-leaf notebooks. Late in the evenings, I would try to imagine pretty hypotheses that tied all of radiation biology together, but so much special pleading was necessary that I almost never tried to explain them to Luria, much less to Delbrück.

In the early fall the question came up where I should go once I got my Ph.D. Europe seemed the natural place since, in the Luria-Delbrück circle, the constant reference to their early lives left me with the unmistakable feeling that Europe's slower paced traditions were more conducive to the production of first-rate ideas. I was thus urged to go to Herman Kalckar's lab in Copenhagen, since in 1946 he had taken the phage course and now professed to want to study phage reproduction. Though Kalckar was admittedly a biochemist, through his brother he knew Bohr and so should be receptive to the need of high powered theoretical reasoning. Even better, Kalckar's interest in nucleotide chemistry should immediately be applicable to the collection of nucleotides in DNA. The decision was finally settled in early November, when by accident Kalckar and Delbrück both were in Chicago on a weekend when Szilard had got the midwestern phage people together for a small meeting. Kalckar seemed excited about the possibility of using some C^{14} labeled adenine, which had just been synthesized in Copenhagen, to study phage reproduction, and he gave the impression of very much wanting phage people to come to his lab.

The midwestern phage meetings were then being held almost every month in Chicago, thanks to a small grant to Szilard from the Rockefeller Foundation, which covered some of the travel expenses and all of the food bills. Lederberg also began to come, adding a new vocal dimension, for he could give non-stop 3-4 hour orations without making a dent in the experiments he thought we should know about. By then, he and his wife had found phage λ in E. coli strain K12 but perhaps because of Delbrück's dislike of the possibility of lysogeny, I paid little attention to the discovery. Instead I conserved my brain for the facts about the somewhat messy partial diploids. My guess is that no one left the meetings remembering more than a small fraction of the ingenious alternative explanations that Lederberg dreamed up to explain the increasing number of paradoxes arising from his experiments.

In the spring of 1950 Luria went back to the problem of the distribution of spontaneous mutations within single bursts of infected cells, hoping he would find out whether or not the genetic material replicated exponentially. I spent a month on the first version of my thesis, but Luria did not like it and took it home for rewriting. This left me little more to do and not surprisingly the thesis was accepted without fuss at my Ph.D. exam in late May. Then I went out to Pasadena for a month, flying back East to spend a final six weeks in Cold Spring Harbor before the boat would take me to Europe. For a brief while I was afraid that the outbreak of the Korean War might keep me from sailing but without hesitation my draft board gave me permission to leave the country.

Practical jokes dominated the mood during the late summer in Cold Spring Harbor, culminating in an evening when Gordon Lark, Victor Bruce, and his sister Manny Delbrück, and I let the air out of the tires of several friends' cars parked before Neptune's Cave. Afterwards buckets of water were poured over our beds. On another evening, Visconti interrupted a staid Demerec social evening with an attack with a toy machine gun.

The growing number of phage people became noticeable at the phage meeting in late August. Some thirty people came, I being most affected by the talk of Kozloff and Putnam on their failure to observe 100% transfer of parental phage P^{32} to the progeny particles. Instead they believed that only somewhere between 20% and 40% of the parental label was transferred. While there existed loopholes in their experiments, the possibility was raised that perhaps both genetic and non-genetic phage DNA existed and that only the genetic portion was transferred. That prompted Seymour Cohen to predict that a second generation of growth might yield 100% transfer.

These ideas I followed up as soon as I got to Europe. Gunther Stent had also chosen Copenhagen, and so Kalckar was faced with two phage people far less interested in biochemistry than he had been led to expect. At the same time, when we could follow Kalckar's words, it was apparent that he was not fixed on the problem of gene replication and seemed happier talking about nucleoside rearrangements. Luckily Kalchar's close friend, Ole Maaløe, had been bitten with the phage bug, and without ever formally acknowledging the arrangement, Stent and I began working with Maaløe in his lab at the State Serum Institut. Maaløe liked the idea of the second generation experiment and we began making labeled phage. After a few failures, we obtained the clean cut, but then disappointing result that the transfer in the second generation was the same as in the first generation. The data were quickly written up and dispatched in early February

(1951) to Delbrück for his approval and possible transmission to the Proceedings of the National Academy. My turgid style was quickly rejected by Delbrück, who completely rewrote the introduction and discussion sections and then sent it on.

By then I knew that Maaløe wanted to go to Caltech the following autumn and so I had to find a place for the next year. I thus wrote to Luria of my dilemma, indicating a preference for England and mentioning Bawden and Pirie, neither of whom I had met. In Luria's reply he took me to task for laziness, saying that I should use my time to acquire the physics and chemistry necessary for a real breakthrough. Clearly my Copenhagen period was not developing the way Luria wanted it. Instead of learning anything new, Stent and I were merely transferring the phage group spirit to Denmark. The net result would be that I would end up doing routine phage work, and if that were to be the case it would make better sense for me to be in the States.

Some months later, Luria responded more warmly when I suggested that I go to Perutz's group at the Cavendish Laboratory, to work on the structures of DNA and the plant viruses. Soon after my letter came, he met John Kendrew at Ann Arbor and set into motion the events which led me to Cambridge and Francis Crick. How the DNA structure fell out I shall not tell here since the story is involved, and is soon to be published elsewhere.

A glossary of selected names and places mentioned by James D. Watson in *Growing up in the Phage Group.*

Adams, Mark.   See *Cold Spring Harbor*

Bacteriophage.   (Phage for short.) A virus capable of infecting and destroying bacterial cells. About twenty minutes after the infection of one bacteruim by a single phage particle, the bacterial cell bursts and releases from 100 to 200 progeny phage particles into the surrounding medium. (The "d'Herelle substances" of the first essay of this section were, in fact, bacteriophage.)

Cold Spring Harbor   A small village on the North Shore of Long Island near Oyster Bay where there were two research laboratories (now joined into a single one): the Biological Laboratory and the Department of Genetics of the Carnegie Institution of Washington (D.C.). Many of the names mentioned by Watson are those of scientists who were in residence at these laboratories:

    Adams, Mark.   A phage geneticist who for many years taught a summer course on bacteriophage techniques at Cold Spring Harbor. This course was the reason that phage workers from all parts of the world congregated at the Biological Laboratory. Many physicists who turned to biology following World War II turned to phage and learned their biological ABC's in Mark Adam's phage course.

    Demerec, Milislav.   The director of the two laboratories during the time

Watson described. Demerec's studies in bacterial genetics revealed that the genes controlling several steps in a metabolic pathway are physically arranged in the same sequence on the bacterial chromosome as are the chemical steps in the metabolic pathway they control; this is a pioneer study of structure and function at the nuclear level.

Doermann, August (Gus). In 1948, a young microbial geneticist working in Demerec's laboratory.

Hershey, Alfred A phage geneticist who, by labeling the protein coats of one lot of phage particles with radioactive sulfur and the DNA of another lot with radioactive phosphorus, showed that only the DNA of a virus particle enters the bacterium as an infective agent; the protein coat remains outside the bacterium and is discarded. In 1969, Hershey shared the Nobel prize with Luria and Delbrück.

Kelner, Albert. In 1948, a young microbial geneticist working at the Biological Laboratory.

McClintock, Barbara. One of the world's leading cytogeneticists who, in working with mutant strains of field corn (maize), discovered and unraveled the interactions of genes that lead to the regulation of one gene's action by a second one; this discovery anticipated many of the control systems now known in microorganisms. Dr. McClintock was awarded the National Medal of Science in 1970.

Visconti, Nicholas. In 1950, a young microbial geneticist working in Demerec's laboratory. (Nick is the youngest of the well-known Visconti brothers of Italy, one of whom is Luchino, the motion picture director.)

Delbrück, Max. A physicist with a long-standing interest in biological problems. Together with Hershey and Luria he laid the ground work for research in phage genetics; the three were awarded the Nobel prize in 1969.

Demerec, Milislav. See Cold Spring Harbor.

Doermann, August (Gus). See Cold Spring Harbor.

Dulbecco, Renato. One of Luria's colleagues who, after mastering the techniques needed for the study of bacteriophage, turned to a study of viruses in mammalian tissue culture cells and, more recently, the role of viruses in the origin of cancer.

Hershey, Alfred. See Cold Spring Harbor.

Kelner, Albert. See Cold Spring Harbor.

Lederberg, Joshua. A microbial geneticist who, while still a graduate student, discovered a sexual process in bacteria (the colon bacillus, *Escherichia coli*) that made these organisms unexcelled for genetic studies. Together with George Beadle and Edward Tatum, he was awarded the Nobel prize in 1958.

Luria, Salvadore.   One of the early founders of phage genetics; awarded the Nobel prize in 1969.

McClintock, Barbara.   See Cold Spring Harbor.

Michaelis, Leonor.   A long-time summer visitor to the Biological Laboratory at Cold Spring Harbor. Michaelis was a pioneer biochemist who specialized in rates of enzymatic reactions; he was not an active research worker at Cold Spring Harbor during the summers described by Watson.

Muller, H.J.   An early geneticist of the T.H. Morgan group at Columbia University; this was the group that made the common vinegar fly (*Drosophila melanogaster*) into the leading experimental material for genetic studies – a position it held for twenty years or more. Muller was awarded the Nobel prize in 1946 for his work on radiation genetics, including the initial demonstration that radiation causes gene mutations.

Novick, Aaron.   One of the physicists who entered biology after World War II; he was a close colleague of Leo Szilard – indeed, among early phage workers the two names very nearly blended into one: Novick'nz'lard.

Phage.   See Bacteriophage.

Szilard, Leo.   A physicist who (through Albert Einstein) was instrumental in persuading President Roosevelt that the United States should develop the atomic bomb during World War II. After the war he and many other theoretical physicists switched their interests to biology, and, in particular, to phage genetics.

Visconti, Nicholas.   See Cold Spring Harbor.

The Double Helix

James D. Watson

In the summer of 1955, I arranged to join some friends who were going into the Alps. Alfred Tissieres, then a Fellow at King's, had said he would get me to the top of the Rothorn, and even though I panic at voids this did not seem to be the time to be a coward. So after getting in shape by letting a guide lead me up the Allinin, I took the two-hour postal-bus trip to Zinal, hoping that the driver was not carsick as he lurched the bus around the narrow road twisting above the falling rock slopes. Then I saw Alfred standing in front of the hotel, talking with a long-mustached Trinity don who had been in India during the war.

Since Alfred was still out of training, we decided to spend the afternoon walking up to a small restaurant which lay at the base of the huge glacier falling down off the Obergabelhorn and over which we were to walk the next day. We were only a few minutes out of sight of the hotel when we saw a party coming down upon us, and I quickly recognized one of the climbers. He was Willy Seeds, a scientist who several years before had worked at King's College, London, with Maurice Wilkins on the optical properties of DNA fibers. Willy soon spotted me, slowed down, and momentarily gave the impression that he might remove his rucksack and chat for a while. But all he said was, "How's Honest Jim?" and quickly increasing his pace was soon below me on the path.

Later as I trudged upward, I thought again about our earlier meetings in London. Then DNA was still a mystery up for grabs, and no one was sure who would get it and whether he would deserve it if it proved as exciting as we semisecretly believed. But now the race was over and, as one of the winners, I knew the tale was not simple and certainly not as the newspapers reported. Chiefly it was a matter of five people: Maurice Wilkins, Rosalind Franklin, Linus Pauling, Francis Crick, and me. And as Francis was the dominant force in shaping my part, I will start the story with him.

1

I have never seen Francis Crick in a modest mood. Perhaps in other company he is that way, but I have never had reason so to judge him. It has nothing to do with his present fame. Already he is much talked about, usually with reverence, and someday he may be considered in the category of Rutherford or Bohr. But this was not true when, in the fall of 1951, I came to the Cavendish Laboratory of Cambridge University to join a small group of physicists and chemists working on the three-dimensional structures of proteins. At that time he was thirty-five, yet almost totally unknown. Although some of his closest colleagues realized the value of his quick, penetrating mind and frequently sought his advice, he was often not appreciated, and most people thought he talked too much.

Leading the unit to which Francis belonged was Max Perutz, an Austrian-born chemist who came to England in 1936. He had been collecting X-ray diffraction data from hemoglobin crystals for over ten years and was just beginning to get somewhere. Helping him was Sir Lawrence Bragg, the director of the Cavendish. For almost forty years Bragg, a Nobel Prize winner and one of the founders of crystallography, had been watching X-ray diffraction methods solve structures of ever-increasing difficulty. The more complex the molecule, the happier Bragg became when a new method allowed its elucidation. Thus in the immediate postwar years he was especially keen about the possibility of solving the structures of proteins, the most complicated of all molecules. Often, when administrative duties permitted, he visited Perutz' office to discuss recently accumulated X-ray data. Then he would return home to see if he could interpret them.

Somewhere between Bragg the theorist and Perutz the experimentalist was Francis, who occasionally did experiments but more often was immersed in the theories for solving protein structures. Often he came up with something novel, would become enormously excited, and immediately tell it to anyone who would listen. A day or so later he would often realize that his theory did not work and return to experiments, until boredom generated a new attack on theory.

There was much drama connected with these ideas. They did a great deal to liven up the atmosphere of the lab, where experiments usually lasted several months to years. This came partly from the volume of Crick's voice: he talked louder and faster than anyone else and, when he laughed, his location within the Cavendish was obvious. Almost everyone enjoyed these manic moments, especially when we had the time to listen attentively and to tell him bluntly when we lost the train of his argument. But there was one notable exception. Conversations with Crick frequently upset Sir Lawrence Bragg, and the sound of his voice was often sufficient to make Bragg move to a safer room. Only infrequently would he come to tea in the Cavendish, since it meant enduring Crick's booming over the tea room. Even then Bragg was not completely safe. On two occasions the corridor outside his office was flooded with water pouring out of a laboratory in which Crick was working. Francis, with his interest in theory, had neglected to fasten securely the rubber tubing around his suction pump.

At the time of my arrival, Francis' theories spread far beyond the confines of protein crystallography. Anything important would attract him, and he frequently visited other labs to see which new experiments had been done. Though he was generally polite and considerate of colleagues who did not realize the real meaning of their latest experiments, he would never hide this fact from them. Almost immediately he would suggest a rash of new experiments that should confirm his interpretation. Moreover, he would not refrain from subsequently telling all who would listen how his clever new idea might set science ahead.

As a result, there existed an unspoken yet real fear of Crick, especially among his contemporaries who had yet to establish their reputations. The quick manner in which he seized their facts and tried to reduce them to coherent patterns frequently made his friends' stomachs sink with the apprehension that, all too often in the near future, he would succeed, and expose to the world the fuzziness of minds hidden from direct view by the considerate, well-spoken manners of the Cambridge colleges.

Though he had dining rights for one meal a week at Caius College, he was not yet a fellow of any college. Partly this was his own choice. Clearly he did not want to be burdened by the unnecessary sight of undergraduate tutees. Also a factor was his laugh, against which many dons would most certainly rebel if subjected to its shattering bang more than once a week. I am sure this occasionally bothered Francis, even though he obviously knew that most High Table life is dominated by pedantic, middle-aged men incapable of either amusing or educating him in anything worthwhile. There always existed King's College, opulently nonconformist and clearly capable of absorbing him without any loss of his or its character. But despite much effort on the part of his friends, who knew he was a delightful dinner companion, they were never able to hide the fact that a stray remark over sherry might bring Francis smack into your life.

2

Before my arrival in Cambridge, Francis only occasionally thought about deoxyribonucleic acid (DNA) and its role in heredity. This was not because he thought it uninteresting. Quite the contrary. A major factor in his leaving physics and developing an interest in biology had been the reading in 1946 of What is Life? by the noted theoretical physicist Erwin Schrödinger. This book very elegantly propounded the belief that genes were the key components of living cells and that to understand what life is, we must know how genes act. When Schrödinger wrote his book (1944), there was general acceptance that genes were special types of protein molecules. But almost at this same time the bacteriologist O. T. Avery was carrying out experiments at the Rockefeller Institute in New York which showed that hereditary traits could be transmitted from one bacterial cell to another by purified DNA molecules.

Given the fact that DNA was known to occur in the chromosomes of all cells, Avery's experiments strongly suggested that future experiments would show that all genes were composed of DNA. If true, this meant to Francis that proteins would not be the Rosetta Stone for unraveling the true secret of life. Instead, DNA would have to provide the key to enable us to find out how the genes determined, among other characteristics, the color of our hair, our eyes, most likely our comparative intelligence, and maybe even our potential to amuse others.

Of course there were scientists who thought the evidence favoring DNA was inconclusive and preferred to believe that genes were protein molecules. Francis, however, did not worry about these skeptics. Many were cantankerous fools who unfailingly backed the wrong horses. One could not be a successful scientist without realizing that, in contrast to the popular conception supported by newspapers and mothers of scientists, a goodly number of scientists are not only narrow-minded and dull, but also just stupid.

Francis, nonetheless, was not then prepared to jump into the DNA world. Its basic importance did not seem sufficient cause by itself to lead him out of the protein field which he had worked in only two years and was just beginning to master intellectually. In addition, his colleagues at the Cavendish were only marginally interested in the nucleic acids, and even in the best of financial circumstances it would take two or three years to set up a new research group primarily devoted to using X rays to look at the DNA structure.

Moreover, such a decision would create an awkward personal situation. At this time molecular work on DNA in England was, for all practical purposes, the personal property of Maurice Wilkins, a bachelor who worked in London at King's College*. Like Francis, Maurice had been a physicist and also used X-ray diffraction as his principal tool of research. It would have looked very bad if Francis had jumped in on a problem that Maurice had worked over for several years. The matter was even worse because the two, almost equal in age, knew each other and, before Francis remarried, had frequently met for lunch or dinner to talk about science.

It would have been much easier if they had been living in different

---

*A division of the University of London, not to be confused with King's College, Cambridge.

countries. The combination of England's coziness — all the important people, if not related by marriage, seemed to know one another — plus the English sense of fair play would not allow Francis to move in on Maurice's problem. In France, where fair play obviously did not exist, these problems would not have arisen. The States also would not have permitted such a situation to develop. One would not expect someone at Berkeley to ignore a first-rate problem merely because someone at Cal Tech had started first. In England, however, it simply would not look right.

Even worse, Maurice continually frustrated Francis by never seeming enthusiastic enough about DNA. He appeared to enjoy slowly understating important arguments. It was not a question of intelligence or common sense. Maurice clearly had both; witness his seizing DNA before almost everyone else. It was that Francis felt he could never get the message over to Maurice that you did not move cautiously when you were holding dynamite like DNA. Moreover, it was increasingly difficult to take Maurice's mind off his assistant, Rosalind Franklin.

Not that he was at all in love with Rosy, as we called her from a distance. Just the opposite — almost from the moment she arrived in Maurice's lab, they began to upset each other. Maurice, a beginner in X-ray diffraction work, wanted some professional help and hoped that Rosy, a trained crystallographer, could speed up his research. Rosy, however, did not see the situation this way. She claimed that she had been given DNA for her own problem and would not think of herself as Maurice's assistant.

I suspect that in the beginning Maurice hoped that Rosy would calm down. Yet mere inspection suggested that she would not easily bend. By choice she did not emphasize her feminine qualities. Though her features were strong, she was not unattractive and might have been quite stunning had she taken even a mild interest in clothes. This she did not. There was never lipstick to contrast with her straight black hair, while at the age of thirty-one her dresses showed all the imagination of English blue-stocking adolescents. So it was quite easy to imagine her the product of an unsatisfied mother who unduly stressed the desirability of professional careers that could save bright girls from marriages to dull men. But this was not the case. Her dedicated, austere life could not be thus explained — she was the daughter of a solidly comfortable, erudite banking family.

Clearly Rosy had to go or be put in her place. The former was obviously preferable because, given her belligerent moods, it would be very difficult for Maurice to maintain a dominant position that would allow him to think unhindered about DNA. Not that at times he didn't see some reason for her complaints — King's had two combination rooms, one for men, the other for women, certainly a thing of the past. But he was not responsible, and it was not pleasure to bear the cross for the added barb that the women's combination room remained dingily pokey whereas money had been spent to make life agreeable for him and his friends when they had their morning coffee.

Unfortunately, Maurice could not see any decent way to give Rosy the boot. To start with, she had been given to think that she had a position for several years. Also, there was no denying she had a good brain. If she could only keep her emotions under control, there would be a good chance that she could really help him. But merely wishing for relations to improve was taking something of a gamble, for Cal Tech's fabulous chemist Linus Pauling was not subject to the confines of British fair play. Sooner or later Linus, who had just turned fifty, was bound to try for the most important of all scientific prizes. There was no doubt that he was interested. Our first principles told us that Pauling could not be the greatest of all chemists without realizing that DNA was the most golden of all molecules. Moreover, there was definite proof. Maurice had received a letter from Linus asking for a copy of the crystalline DNA X-ray photographs. After some hesitation he wrote back saying that he wanted to look more closely at the data before releasing the pictures.

All this was most unsettling to Maurice. He had not escaped into biology only to find it personally as objectionable as physics, with its atomic consequences. The combination of both Linus and Francis breathing down his neck often made it very difficult to sleep. But at least Pauling was six thousand miles away, and even Francis was separated by a two-hour rail journey. The real problem, then, was Rosy. The thought could not be avoided that the best home for a feminist was in another person's lab.

3

It was Wilkins who had first excited me about X-ray work on DNA. This happened at Naples when a small scientific meeting was held on the structures of the large molecules found in living cells. Then it was the spring of 1951, before I knew of Francis Crick's existence. Already I was much involved with DNA, since I was in Europe on a postdoctoral fellowship to learn its biochemistry. My interest in DNA had grown out of a desire, first picked up while a senior in college, to learn what the gene was. Later, in graduate school at Indiana University, it was my hope that the gene might be solved without my learning any chemistry. This wish partially arose from laziness since, as an undergraduate at the University of Chicago, I was principally interested in birds and managed to avoid taking any chemistry or physics courses which looked of even medium difficulty. Briefly the Indiana biochemists encouraged me to learn organic chemistry, but after I used a bunsen burner to warm up some benzene, I was relieved from further true chemistry. It was safer to turn out an uneducated Ph.D. than to risk another explosion.

So I was not faced with the prospect of absorbing chemistry until I went to Copenhagen to do my postdoctoral research with the biochemist Herman Kalckar. Journeying abroad initially appeared the perfect solution to the complete lack of chemical facts in my head, a condition at times encouraged by my Ph.D. supervisor, the Italian-trained microbiologist Salvador Luria. He positively abhorred most chemists, especially the competitive variety out of the jungles of New York City. Kalckar, however, was obviously cultivated, and Luria hoped that in his civilized, continental company I would learn the necessary tools to do chemical research, without needing to react against the profit-oriented organic chemists.

Then Luria's experiments largely dealt with the multiplication of bacterial viruses (bacteriophages, or phages for short). For some years the suspicion had existed among the more inspired geneticists that viruses where a form of naked genes. If so, the best way to find out what a gene was and how it duplicated was to study the properties of viruses. Thus, as the simplest viruses were the phages, there had sprung up between 1940 and 1950 a growing number of scientists (the phage group) who studied phages with the hope that they would eventually learn how the genes controlled cellular heredity. Leading this group were Luria and his German-born friend, the theoretical physicist Max Delbrück, then a professor at Cal Tech. While Delbrück kept hoping that purely genetic tricks could solve the problem, Luria more often wondered whether the real answer would come only after the chemical structure of a virus (gene) had been cracked open. Deep down he knew that it is impossible to describe the behavior of something when you don't know what it is. Thus, knowing he could never bring himself to learn chemistry, Luria felt the wisest course was to send me, his first serious student, to a chemist.

He had no difficulty deciding between a protein chemist and a nucleic-acid chemist. Though only about one half the mass of a bacterial virus was DNA (the other half being protein), Avery's experiment made it smell like the essential genetic material. So working out DNA's chemical structure might be the essential step in learning how genes duplicated. Nonetheless, in contrast to the proteins, the solid chemical facts known about DNA were meager. Only a few chemists worked with it and, except for the fact that nucleic acids were very large molecules built up from smaller building blocks, the nucleotides, there was almost nothing chemical that the geneticist could grasp at. Moreover, the chemists who did work on DNA were almost always organic chemists with no interest in genetics. Kalckar was a bright exception. In the summer of 1945 he had come to the lab at Cold Spring Harbor, New York, to take Delbrück's course on bacterial viruses. Thus both Luria and Delbrück hoped the Copenhagen lab would be the place where the combined techniques of chemistry and genetics might eventually yield real biological dividends.

Their plan, however, was a complete flop. Herman did not stimulate me in the slightest. I found myself just as indifferent to nucleic-acid chemistry in his lab as I had been in the States. This was partly because I could not see how the type of problem on which he was then working (metabolism of nucleotides) would lead to anything of immediate interest to genetics. There was also the fact that, though Herman was obviously civilized, it was impossible to understand him.

I was able, however, to follow the English of Herman's close friend Ole Maaløe. Ole had just returned from the States (Cal Tech), where he had become very excited about the same phages on which I had worked for my degree. Upon his return he gave up his previous research problem and was devoting full time to phage. Then he was the only Dane working with phage and so was quite pleased that I and Gunther Stent, a phage worker from Delbrück's lab, had come to do research with Herman. Soon Gunther and I found ourselves going regularly to visit Ole's lab, located several miles from Herman's, and within several weeks we were both actively doing experiments with Ole.

At first I occasionally felt ill at ease doing conventional phage work with Ole, since my fellowship was explicitly awarded to enable me to learn biochemistry with Herman; in a strictly literal sense I was violating its terms. Moreover, less than three months after my arrival in Copenhagen I was asked to propose plans for the following year. This was no simple matter, for I had no plans. The only safe course was to ask for funds to spend another year with Herman. It would have been risky to say that I could not make myself enjoy biochemistry. Furthermore, I could see no reason why they should not permit me to change my plans after the renewal was granted. I thus wrote to Washington saying that I wished to remain in the stimulating environment of Copenhagen. As expected, my fellowship was then renewed. It made sense to let Kalckar (whom several of the fellowship electors knew personally) train another biochemist.

There was also the question of Herman's feelings. Perhaps he minded the fact that I was only seldom around. True, he appeared very vague about most things and might not yet have really noticed. Fortunately, however, these fears never had time to develop seriously. Through a completely unanticipated event my moral conscience became clear. One day early in December, I cycled over to Herman's lab expecting another charming yet totally incomprehensible conversation. This time, however, I found Herman could be understood. He had something important to let out: his marriage was over, and he hoped to obtain a divorce. This fact was soon no secret — everyone else in the lab was also told. Within a few days it became apparent that Herman's mind was not going to concentrate on science for some time, for perhaps as long as I would remain in Copenhagen. So the fact that he did not have to teach me nucleic-acid biochemistry was obviously a godsend. I could cycle each day over to Ole's lab, knowing it was clearly better to deceive the fellowship electors about where I was working than to force Herman to talk about biochemistry.

At times, moreover, I was quite pleased with my current experiments on bacterial viruses. Within three months Ole and I had finished a set of experiments on the fate of a bacterial-virus particle when it multiplies inside a bacterium to form several hundred new virus particles. There were enough data for a respectable publication and, using ordinary standards, I knew I could stop work for the rest of the year without being judged unproductive. On the other hand, it was equally obvious that I had not done anything which was going to tell us what a gene was or how it reproduced. And unless I became a chemist, I could not see how I would.

I thus welcomed Herman's suggestion that I go that spring to the Zoological Station at Naples, where he had decided to spend the months of April and May. A trip to Naples made great sense. There was no point in doing nothing in Copenhagen, where spring does not exist. On the other hand, the sun of Naples might be conducive to learning something about the biochemistry of the embryonic development of marine animals. It might also be a place where I could quietly read genetics. And when I was tired of it, I might conceivably pick up a biochemistry text. Without any hesitation I wrote to the States requesting permission to accompany Herman to Naples. A cheerful affirmative letter wishing me a pleasant journey came by return post from Washington. Moreover, it enclosed a $200 check for travel expenses. It made me feel slightly dishonest as I set off for the sun.

4

Maurice Wilkins also had not come to Naples for serious science. The trip from London was an unexpected gift from his boss, Professor J. T. Randall. Originally Randall had been scheduled to come to the meeting on macro-molecules and give a paper about the work going on in his new biophysics lab. Finding himself overcommitted, he had decided to send Maurice instead. If no one went, it would look bad for his King's College lab. Lots of scarce Treasury money had to be committed to set up his biophysics show, and suspicions existed that this was money down the drain.

No one was expected to prepare an elaborate talk for Italian meetings like this one. Such gatherings routinely brought together a small number of invited guests who did not understand Italian and a large number of Italians, almost none of whom understood rapidly spoken English, the only language common to the visitors. The high point of each meeting was the day-long excursion to some scenic house or temple. Thus there was seldom chance for anything but banal remarks.

By the time Maurice arrived I was noticeably restless and impatient to return north. Herman had completely misled me. For the first six weeks in

Naples I was constantly cold. The official temperature is often much less relevant than the absence of central heating. Neither the Zoological Station nor my decaying room atop a six-story nineteenth-century house had any heat. If I had even the slightest interest in marine animals, I would have done experiments. Moving about doing experiments is much warmer than sitting in the library with one's feet on a table. At times I stood about nervously while Herman went through the motions of a biochemist, and on several days I even understood what he said. It made no difference, however, whether or not I followed the argument. Genes were never at the center, or even at the periphery, of his thoughts.

Most of my time I spent walking the streets or reading journal articles from the early days of genetics. Sometimes I daydreamed about discovering the secret of the gene, but not once did I have the faintest trace of a respectable idea. It was thus difficult to avoid the disquieting thought that I was not accomplishing anything. Knowing that I had not come to Naples for work did not make me feel better.

I retained a slight hope that I might profit from the meeting on the structures of biological macromolecules. Though I knew nothing about the X-ray diffraction techniques that dominated structural analysis, I was optimistic that the spoken arguments would be more comprehensible than the journal articles, which passed over my head. I was specially interested to hear the talk on nucleic acids to be given by Randall. At that time almost nothing was published about the possible three-dimensional configurations of a nucleic-acid molecule. Conceivably this fact affected my casual pursuit of chemistry. For why should I get excited learning boring chemical facts as long as the chemists never provided anything incisive about the nucleic acids?

The odds, however, were against any real revelation then. Much of the talk about the three-dimensional structure of proteins and nucleic acids was hot air. Though this work had been going on for over fifteen years, most if not all of the facts were soft. Ideas put forward with conviction were likely to be the products of wild crystallographers who delighted in being in a field where their ideas could not be easily disproved. Thus, although virtually all biochemists, including Herman, were unable to understand the arguments of the X-ray people, there was little uneasiness. It made no sense to learn complicated mathematical methods in order to follow baloney. As a result, none of my teachers had ever considered the possibility that I might do postdoctoral research with an X-ray crystallographer.

Maurice, however, did not disappoint me. The fact that he was a substitute for Randall made no difference: I had not known about either. His talk was far from vacuous and stood out sharply from the rest, several of which bore no connection to the purpose of the meeting. Fortunately these were in Italian, and so the obvious boredom of the foreign guests did not need to be construed as impoliteness. Several other speakers were continental biologists, at that time guests at the Zoological Station, who only briefly alluded to macromolecular structure. In contrast, Maurice's X-ray diffraction picture of DNA was to the point. It was flicked on the screen near the end of his talk. Maurice's dry English form did not permit enthusiasm as he stated that the picture showed much more detail than previous pictures and could, in fact, be considered as arising from a crystalline substance. And when the structure of DNA was known, we might be in a better position to understand how genes work.

Suddenly I was excited about chemistry. Before Maurice's talk I had worried about the possibility that the gene might be fantastically irregular. Now, however, I knew that genes could crystallize; hence they must have a regular structure that could be solved in a straightforward fashion. Immediately I began to wonder whether it would be possible for me to join Wilkins in working on DNA. After the lecture I tried to seek him out. Perhaps he already knew more than his talk had indicated — often if a scientist is not absolutely sure he is correct, he is hesitant to speak in public. But there was no opportunity to talk to him; Maurice had vanished.

Not until the next day, when all the participants took an excursion to the Greek temples at Paestum, did I get an opportunity to introduce myself. While waiting for the bus I started a conversation and explained how interested I was in DNA. But before I could pump Maurice we had to board, and I joined my sister, Elizabeth, who had just come in from the States. At the temples we all scattered, and before I could corner Maurice again I realized that I might have had a tremendous stroke of good luck. Maurice had noticed that my sister was very pretty, and soon they were eating lunch together. I was immensely pleased. For years I had sullenly watched Elizabeth being pursued by a series of dull nitwits. Suddenly the possibility opened up that her way of life could be changed. No longer did I have to face the certainty that she would end up with a mental defective. Furthermore, if Maurice really liked my sister, it was inevitable that I would become closely associated with his X-ray work on DNA. The fact that Maurice excused himself to go and sit alone did not upset me. He obviously had good manners and assumed that I wished to converse with Elizabeth.

As soon as we reached Naples, however, my daydreams of glory by association ended. Maurice moved off to his hotel with only a casual nod. Neither the beauty of my sister nor my intense interest in the DNA structure had snared him. Our futures did not seem to be in London. Thus I set off to Copenhagen and the prospect of more biochemistry to avoid.

DNA: Blueprint for Life:
An Essay in Three Parts

1. Self-replication as the basis for life

The casual observer can rather easily classify nearby objects as living and nonliving. Animals and plants are living; metal objects, rocks, books, and lamp shades are not. Even when pushed beyond his powers of direct observation, our observer will still assert that bacteria and other microorganisms are living. He might, too, admit that the "heartwood" of a tree or such things as his own hair and fingernails are not living but, in fact, are dead.

The success of our observer in reaching a passable classification resembles that of biologists and natural philosophers over many centuries. Life, as a property of living things, has certain characteristics; things are said to be alive if they are composed, in part at least, of a special watery, proteinaceous substance called *protoplasm*. Living things have the ability to ingest substances (living or nonliving) from their immediate surroundings and to convert these substances into additional living material of their own sort; they grow. They also reproduce by any one of several mechanisms to produce more of their own kind; the new individuals in turn are able to grow and reproduce. Many living things are irritable; they possess the power of movement and, especially in the presence of substances or circumstances harmful to life, withdraw from dangerous or otherwise unpleasant situations.

Simple classification schemes generally prove to be inadequate when pressed far enough; the simple view of life and living things that I have just outlined breaks down when we discuss forms of life smaller than bacteria. Viruses, for example, can be obtained in crystalline form. Some viruses are composed of but one or two substances. It is tempting to deny these simple organisms the dignity of life but to do so requires that we spell out in considerable detail and justify those properties we deem essential for life in which viruses are lacking. I prefer to adopt the easier alternative, that of including viruses among living organisms. In this case, the definition of life must be made correspondingly simple; a living organism is one that possesses the ability to direct under favorable circumstances the synthesis of more of its own kind. As a practical matter we might point out that the rate at which new individuals are or may be produced under favorable circumstances must exceed the overall rate at which similar individuals disintegrate or otherwise cease to function. The stipulation about rates of formation and rates of decay must apply to forms living today; unstable, simple systems that may have existed at one time, even though they may have been capable of self-replication, would not have persisted if their rate of decay exceeded their rate of successful replication.

What are the essential features of a self-directing, self-replicating structure? At the very simplest, two structures are needed; these were hinted at with remarkable foresight by Professor H. J. Muller:*

> We must note further that, in all such cases in ordinary chemistry, if we caused a change in the composition of the autocatalytic [=self-replicating] substance we should practically always render it non-autocatalytic. For the change will either leave the substance still capable of affecting the original reaction in the same way as before – in which case the substance will not be causing the production of material of exactly its own kind any more – or, if the effect of the substance on the original reaction becomes altered, it would be most remarkable if it chanced to be altered in precisely such a direction that the reaction now caused was of just such a nature as to produce the substance in the exact form in which it had become modified. Such reciprocity between the change in the specifically autocatalytic substance and the change in the effect which it had on the substrate would require a special kind of construction on the former, of a sort not yet known to the chemists, which allowed of variation in certain features of its pattern while at the same time a mechanism was retained which caused the variable pattern to be copied in its present form by the substances newly formed. A specific autocatalytic substance of this kind would fulfill our definition of a gene since it would be able to "mutate" without losing its property of "propagation." If able to mutate thus, indefinitely, it would eventually go through an evolution akin to biological evolution.

And later he wrote:

> What the structure of the gene itself actually was, in physico-chemical terms, the modern geneticist would like to know – perhaps beyond all other questions of genetics – but as yet he remains in almost complete ignorance of this matter. What feature or features of its structure allowed it to *mutate* without losing this specific autocatalytic ability can only be guessed at most inadequately now, but in these features lay all the promise of life, as distinguished from the inanimate. The mutations must have been, must still be, rearrangements in a pattern of one sort which leave unchanged certain other arrangements, of an entirely different sort, which are responsible for the specific autocatalysis. The latter, *stable* arrangements somehow result in the copying of the former, mutable arrangements (as well as their own), by the raw material as it becomes organized.

Muller clearly saw that self-replication is the essence of what we call "life."

*From *Studies in Genetics: The Selected Papers of H. J. Muller.* Copyright © 1962 by Indiana University Press. Reprinted by permission of the publisher.

He also saw that self-replication is a property of genes, the basic units of heredity. The complicated forms of higher organisms and, indeed, the complex nature of the protoplasm of even the simplest forms of life are to be explained as by-products of replicating genes rather than the reverse. The reverse possibility, that self-replicating genes are a by-product of "living" protoplasm, was discarded by Muller because the "living protoplasm" would still require self-replicating particles for its continued existence. Convinced by the logic of these deductions, Muller predicted the gene-like nature of virus particles (known then as "d'Herelle substances") before their true nature was known; it seemed unlikely to him that two substances — gene substance and d'Herelle substance — should have identical self-replicating properties but, despite this great similarity, have unrelated origins. Later, when it was known that heritable properties could be transferred from one strain of bacteria to another by means of extracted deoxyribonucleic acid (DNA), Muller again saw this as another step toward the day when (as he had predicted much earlier) "we may be able to grind genes in a mortar and cook them in a beaker."

The chemical nature of the gene is now known. DNA is indeed the genetic material, as experiments on the transfer of heritable characteristics from one bacterium to another had suggested. Furthermore, J. D. Watson and F. H. C. Crick were able to show, largely by use of X-ray diffraction analyses, that DNA has a double-stranded structure in which each strand serves as the complement of the other (see the accompanying diagram). DNA is even more cleverly built than Muller had dared imagine. Muller had visualized substance *A* (the gene) forming a complementary model *B* out of the surrounding substrate; *B* then regenerated new genes of substance *A*. If the gene were a mutant, the complementary copy *B* would reflect the mutation and would faithfully rebuild mutant forms of substance *A*. The DNA molecule simplifies this process; *A* and *B* are complementary strands of the same chemical; in most organisms the molecule exists as a double structure composed of the accurately paired complementary strands. Each of these strands is capable of guiding the synthesis of its complement, and so accurate self-replication (that is, the *control* of self-replication; not the *energy* that is required for the synthesis of a new molecule) is a built-in feature of genetic material itself. Furthermore, the mechanism for DNA replication provides for the accurate replication of mutations. In this way, DNA satisfies the requirements specified by geneticists on theoretical grounds before the role and structure of DNA were known.

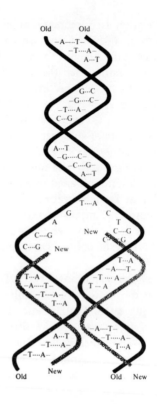

Replication of the DNA molecule. The old helix unwinds (center) and the two new helices are formed. [From Carl P. Swanson, *The Cell,* 3rd ed. (Englewood Cliffs, N.J.: Prentice-Hall, Inc., 1969), p. 56. Reprinted by permission.]

2. The control
of physiological processes

      Just *how* these genes thus determine the reaction-potentialities of the organism and so its resultant form and functioning, is another series of problems, at present practically a closed book in physiology, and one which physiologists as yet seem to have neither the means nor the desire to open!*

The previous essay concentrated on self-replication as a fundamental aspect of living matter; this path led us to DNA — the genetic material for nearly

*From *Studies in Genetics: The Selected Papers of H. J. Muller.* Copyright © 1962 by Indiana University Press. Reprinted by permission of the publisher.

all forms of life, a chemical whose double-stranded, internally complementary structure assures its precise self-replication. It is inconceivable that complex organisms can grow and reproduce without a replicating set of instructions that can be transmitted repeatedly from cell to cell and from parent to offspring. Muller argued on this basis alone that self-replicating gene-like forms of life preceded the more complicated ones. The complicated ones, at any rate, would need the simple ones.

The two functions of genes — self-replication and the control of developmental and physiological processes — were clearly recognized by the early geneticists; the quotation that opens this essay is concerned with the gene's role in the control of physiological processes. The structure of DNA deciphered by Watson and Crick reveals not only how accurate self-replication, in terms of chemical structure, is achieved but also how DNA can assert its influence over biochemical or metabolic reactions. These reactions are mediated by enzymes, organic catalysts in the form of large protein molecules, that are responsible for the occurrence of chemical reactions within cells. The genetic problems associated with the control of these reactions are (1) the synthesis of proteins according to precise structural plans and (2) the synthesis of these enzymes at the proper time, in the proper tissues, and in the proper amounts. The first problem concerns the gene's *ability* to make an enzyme; the second, its ability to *control* enzyme synthesis.

Proteins (of which egg white, hemoglobin, muscle fibers, and enzymes are examples) are long chains composed of relatively simple molecules (*amino acids*); these polypeptide chains are then twisted and folded to produce molecules with characteristic three-dimensional shapes. The complex structure of these molecules is stabilized by chemical cross-linkages between certain amino acids; these linkages correspond to the spot welds that give stability to a modernistic chair that is made by bending and shaping a single long metal rod.

The eventual shape of a protein molecule is specified by the order and the relative positions of amino acids in the long chain; there are 20 of these amino acids in all. Some have charged atoms; some disrupt the linearity of the sequence and force the chain to bend abruptly; some are attracted to water whereas others are repelled by it. These properties and others enable the final molecule to function properly whether its job is the transportation of oxygen atoms from the lung to remote tissues, splitting large ingested molecules into smaller ones, or the synthesis of large molecules from smaller ones as in the liver.

How can a chain of roughly 100 amino acids be put together accurately if there are 20 amino acids that might be inserted into each of the 100 positions on the chain? There is an enormous number of possible chains (20^{100} or, approximately, 10^{130}), of which only one or very few at best are correct for a given job. The problem compares to the need to withdraw from a large box of loose playing cards a sequence specified in advance as: ace of hearts, two of spades, king of clubs, four of clubs, jack of hearts, ten of diamonds, and so forth.

The structure of DNA contains the key to the problem of accurate protein synthesis. We have seen that the replication of DNA must be precise. This precision is mandatory because the matching bases in the complementary strands are either adenosine and thymine or cytosine and guanine. Along either strand, however, the four bases can occur in any order. Whatever the order in which the bases do occur is the order that is faithfully reproduced when the DNA is replicated. The degree to which DNA meets the requirements of genetic material postulated on theoretical grounds by persons such as Professor Muller is truly uncanny.

The order of the bases along a section of DNA determines the order of amino acids along the length of a protein molecule. The actual machinery by which the structure of DNA is translated into the corresponding structure of a protein is complex, but so is the machinery that converts the architect's blueprints into a corresponding building.

The genetic control of protein synthesis is now known as the "central dogma" of molecular biology. For our purposes the following account will suffice. A given amino acid is specified by three adjacent bases of the DNA molecule. Sequential triads of bases spell out (literally; the code has been broken) the eventual sequence of amino acids in the protein molecule. Protein is not made directly from DNA however; rather, the order of the bases on a segment of DNA is first transcribed onto a molecule of related nucleic acid (ribonucleic acid or RNA — specifically, messenger-RNA), which then leaves the nucleus of the cell and enters the cytoplasm. The amino acids are assembled in proper order to make the protein molecule by machinery that "reads" the sequence of bases in the messenger-RNA (or mRNA). The course of events can be written:

$$\text{DNA} \longrightarrow \text{mRNA} \longrightarrow \text{Protein}$$

In any organized system, each part must be able not only to perform but also to restrict its performance to appropriate times. The cellist in an orchestra performs at certain moments, not continuously; a symphony can be identified by the arrangement of musical sounds, and the individual musicians adhere to the controls imposed by that arrangement as interpreted by the conductor.

Genes, too, are subject to various controls. These are understood quite well in some bacteria but less so in higher organisms. The details that are known for each enzyme when multiplied by hundreds of enzymes and the multitude of demands made by different tissues of the body become overwhelming. Once more, it is better to stick to the basics of gene control. The most direct control is one that operates directly between bacterial DNA and the individual's metabolic state. The enzyme that is required to transform substance A into substance B may be synthesized only in the presence of $A;$ that is, the substrate "turns on" the machinery needed for its own metabolism. Substance $B,$ in turn, may be transformed through a series of other enzymatically controlled reactions into substance *F;* the accumulation of *F* in some instances proves to be the event that turns off the gene responsible for the synthesis of the very first enzyme of the metabolic pathway, the one that transformed *A* into *B.* Within each bacterial cell and within each cell of higher organisms there exist networks of interconnecting chemical signaling devices that advise the cell's nucleus of the moment-by-moment state of biochemical and physiological processes. An accumulation of one substance calls for enzymes (perhaps for a whole battery of enzymes concerned with a number of sequential steps in the modification of the given substance) that can transform it into other substances that can be stored or excreted. A deficiency of a substance needed to carry on a vital process calls for enzymes capable of synthesizing it from stored products or from nutrients that diffuse into the cell from the outside.

Many of the control systems operate within the cytoplasm of the cell, well-removed from the nucleus, the chromosomes, and the DNA. Nevertheless, the entire system exists by virtue of the genetic material within which the development and organization of the individual has been outlined through eons of evolution. Like begets like because of the autocatalytic nature of DNA; species differ because changes in DNA, changes that alter either the composition of protein molecules or the control of their production, retain the ability to replicate themselves. The changed forms of DNA in turn control the different times at which enzymes are called into action in different organisms and the various paths that the development of different species takes.

3. Heredity
and individual development

The logical bases of genetics and life were the subjects of the two preceding essays. These necessities, self-reproduction of the gene and its ability to direct physiological processes, were deduced on theoretical grounds by early geneticists (in the 1920's); the physical substance with properties that satisfy these necessities has now been identified as DNA.

The everyday encounters between individuals and genetics occur in the form not of abstract thoughts on life and autocatalytic reactions but of casual wonder over family resemblances or concern over an inherited defect. Montaigne had for a year and a half suffered excruciating recurrent pains from kidney stones when he wrote "Of the Resemblance of Children to Fathers" (1579-80):*

> There is a certain type of subtle humility that is born of presumption, like this one: that we acknowledge our ignorance in many things, and are so courteous as to admit that there are in the works of nature certain qualities and conditions that are imperceptible to us and whose means and causes our capacity cannot discover. By this honest and conscientious declaration we hope to gain credence also about those things that we claim to understand. We have no need to go picking out miracles and remote difficulties; it seems to me that among the things we see ordinarily there are wonders so incomprehensible that they surpass even miracles in obscurity. What a prodigy it is that the drop of seed from which we are produced bears in itself the impressions not only of bodily form but of the thoughts and inclinations of our fathers! Where does that drop of fluid lodge this infinite number of forms? And how do they convey these resemblances with so heedless and irregular a course that the great-grandson will correspond to his great-grandfather, the nephew to the uncle?
>
> In the family of Lepidus in Rome there were three, not in a row but at intervals, who were born with the same eye covered with cartilage. At Thebes there was a family that from their mother's womb bore the figure of a lance-head, and whoever did not bear it was considered illegitimate. Aristotle says that in a certain nation where the women were in common they assigned the children to their fathers by resemblance.
>
> It is probable that I owe this stony propensity to my father, for he died extraordinarily afflicted with a large stone he had in his bladder. He did not perceive his disease until his sixty-seventh year, and before that he had had no threat or symptom of it, in his loins, his sides, or elsewhere. And he had lived until then in a happy state of health, and very little subject to diseases; and he lasted seven years more with this ailment, painfully dragging out the last years of his life. I was born twenty-five years and more before his illness, at a time when he enjoyed his best health, the third of his children in order of birth.
>
> Where was the propensity of this infirmity hatching all this time? And when he was so far from the ailment, how did this slight bit of his substance, with which he made me, bear so great an impression of it for its share? And moreover, how did it remain so concealed that I began to feel it forty-five years later, the only one to this hour out of so many brothers and sisters, and all of the same mother? If anyone will enlighten me about this process, I will believe him about as many other miracles as he wants; provided he does not palm off on me some explanation much more difficult and fantastic than the thing itself.

In this passage, Montaigne raises two, perhaps three, questions that are extremely pertinent to genetics. First, he asks about the transmission of traits, both physical and mental, often with seemingly unaccountable skips, over successive generations. Second, he asks how the drop of (seminal) fluid manages to contain an infinite number of forms; had he known that these forms are lodged in each of millions of spermatozoa contained in a single drop of seminal fluid, he might well have been overwhelmed with incredulity. The existence of spermatozoa and their role in fertilization were not known, however, for two-and-a-half to three centuries after Montaigne's remarks. The third question, that bearing on the transmission of traits not yet apparent in the father at the time of his son's birth, is not a real one; when one understands how an infinite number of forms can be stored within a single spermatozoa (or egg), one understands the principle – not necessarily the details – of individual develop- ment, including the occurrence of specific ailments at specific times of life. In short, if we understand that the unraveling of development from conception to death depends upon the interaction of information encoded in threads of DNA and the prevailing environmental conditions within which the developing individual exists, then we understand how children in their old age can come to resemble their parents at comparable ages. Montaigne's third question is not independent of his second.

The patterns of transmission of hereditary traits from one generation to the next were the heart of genetics as a science from 1900 until the 1920's. Mendelian genetics, of which Mendelian ratios are important features, is largely concerned with this matter. These are the matters that fall within the domain of cytogenetics: the arrangement of genes on chromosomes, the apportionment of chromosomes during the formation of gametes, and the random union of sperm and eggs at the time of fertilization. These are matters that are treated rather well in most modern introductory (including high-school) biology courses.

The problems of development – how a drop of fluid can contain so many forms – are still, four centuries after Montaigne, unresolved in detail. The matter *is* resolved in principle, however, because development must be controlled by the production of certain enzymes in certain amounts at certain times and in certain of the body's cells and tissues; the specification for constructing enzyme molecules and provisions for their production at given times are written in the DNA thread. The blueprints, as we have called them in this essay, are opened and read in the context of the environment. In mammals and birds, the immediate internal environment of the cell is largely under the control of the individual. All organisms control their internal environment, of course, otherwise they would not be alive; mammals and birds, unlike reptiles and lower forms, control the temperature at which internal processes occur. In the final analysis, the control of the internal environment is dependent upon the individual's genetic endowment. The need or lack of need for detailed aspects of internal control is dependent upon the external environment – for example, upon one's diet. Thus, if wheat flour proves to be the undoing of those who are allergic to it, other more acceptable sources of starch in the diet must be found.

Of all the developmental control systems in man, that with the most universal interest is surely sex control, the determination of the sex of the developing individual. This control, much as we might like to think otherwise, is in the hands of destiny just as is the toss of a coin or the throw of a die. The alternative choices arise through the production of two types of sperm by the male. One-half of all sperm carry what is called an X-chromosome, that is, a chromosome bearing a particular set of genes, among which are those whose malfunctioning cause hemophilia and color-blindness. The other half of all sperm carry what is called the Y-chromosome, a chromosome remarkably free of genes with known effects other than sex-determination (in man and other mammals).

In man the alternative choices for sex determination reside with sperm because the eggs carry only X-chromosomes. Consequently, at fertilization an X-bearing egg is fertilized either by an X-bearing sperm to give rise to an XX (female) zygote or by a Y-bearing sperm to give rise to an XY (male) zygote. These contrasting zygotic constitutions – XX and XY – are the reasons why women can produce only X-bearing eggs whereas men produce two types of sperm, those with an X- and those with a Y-chromosome in equal numbers.

A small fraction of all persons have abnormal sexual characteristics that are caused by the abnormal distribution of X- and Y-chromosomes during the formation of gametes. Ordinarily, these chromosomes separate so that one X goes to each egg (in females) or an X or a Y goes to each sperm (in males). Occasionally, XX and O eggs arise because both X-chromosomes or none at all (the symbol, O, is used to designate the absence of a chromosome that otherwise should be present) go to a given egg. Upon fertilization by normal sperm these eggs give rise to XXX, XXY, XO, or YO zygotes. The YO types die; XO individuals are female-like but suffer a number of characteristic physical defects including nonfunctional ovaries; XXY individuals are male-like but have underdeveloped testes and occasionally some breast development; XXX individuals are physically and mentally abnormal females.

Because of the faulty separation of X- and Y-chromosomes, sperm, too, may at times carry the abnormal chromosomal combinations, XY and O. When these sperm fertilize normal X-bearing eggs, XXY and XO zygotes are produced. These individuals are comparable to those with similar chromosomal constitutions described above.

Our lengthy essay on genetics may be terminated now by stating that chromosomes other than X and Y can also be apportioned abnormally during human reproduction. The best known instance is that which is responsible for Down's syndrome (mongolism); individuals suffering the severe mental and physical defects of mongolism carry three rather than two representatives of one very small chromosome. The gross defects of these persons illustrate in a striking manner the delicate balance between different bits of genetic material in their control of the individual's development. Indeed, the apparent lack of similar abnormalities for other, larger chromosomes is presumably caused by the early death of the affected embryos, not by the failure of these chromosomes to disjoin inaccurately during the formation of germ cells.

The Central Dogma:
The Preservation of Simplicity

The central dogma of molecular genetics is a condensation of modern theories of gene action into a single concept that covers both gene replication and protein synthesis. In its most condensed form, the dogma says that DNA is self-replicating and that the sequence of base pairs in the DNA molecule determines the sequence of amino acids in the protein molecule. Ordinarily, the dogma appears in the following diagrammatic form:

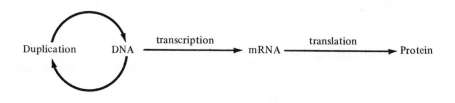

This diagram contains the additional information that DNA controls the structure of a protein molecule by means of an intermediate molecule of mRNA (messenger-RNA); the sequence of bases in DNA determines the (complementary) sequence of bases in mRNA, which in turn determines the sequence of amino acids in the protein molecule.

The central dogma as presented here is one that, in the hands of a competent experimentalist, generates hypothesis after hypothesis, each of which can be tested by means of the new techniques of biochemistry and experimental biology. More experiments, according to one claim, have been suggested by the dogma of molecular genetics than by any conceptual scheme since Darwin's theory of evolution through natural selection. The success with which hypotheses have been suggested, tested, and confirmed as a consequence of the central dogma is itself strong support for the essential truth of the dogma. The more extensive the network of supporting data that surrounds a general theory, the less likely it becomes that an equally good alternative explanation will be devised.

The central dogma, although hinted at in part by Watson and Crick in their early papers, did not emerge automatically as a consequence of their reports. In the first place, the structure of DNA proposed by Watson and Crick differed from the model presented by standard organic chemistry textbooks; the model had to overcome the mental inertia of some scientists. Second, the complementary structure of the two strands of DNA was not immediately accepted as a means for assuring accurate self-replication. Third, the point-by-point relationship between the linear DNA molecule and the amino acid sequence in a protein molecule was established only after a great deal of clever experimental work.

To my knowledge only one serious attack has been launched against the central dogma. By "serious attack" I mean an expression of doubt supported by logical arguments. Many expressions of faith disputing the dogma have been uttered by those who were brought up to believe that proteins carry out all of life's processes including the control of heredity, but for the most part these expressions of faith have been precisely that, and no more. In sharp contrast have been the criticisms of Dr. Barry Commoner.* These criticisms have appeared in semi-popular publications and, as a result, may have caused some confusion in the public mind over modern genetic research.

This essay is devoted to an analysis of Dr. Commoner's views. I do not believe that they are correct. Since they have been presented in a logical manner, however, they require a correspondingly logical response. From an examination of his views emerges the notion that natural selection operating within any species of organisms maintains the basic truth of the central dogma of molecular genetics.

To understand the scheme Commoner would substitute for the central dogma, it is necessary to understand the term "information" as it is used by molecular geneticists. A recipe contains information for baking a cake; the words in a cookbook are translated into a large round doughy dessert as a result of the cook's activities in the kitchen. A laboratory manual contains information that leads to the production of carbon dioxide from limestone; the information is in the manual and is translated by the student chemist. In the same manner, we say that DNA contains the information for its own replication; the complementary strands separate and each serves as a template for reconstructing a copy of the other. We also say that DNA carries the information for protein synthesis. In this case, however, the information carried by DNA is *transcribed* by the synthesis of an RNA molecule from which the information is *translated* in the making of the protein molecule. Similarly, when a library refuses to release a rare volume that is written in a foreign language, the contents of the book can be *transcribed* by the omnipresent Xerox machine; the Xerox copy then serves as the basis for eventual *translation*.

Commoner's issue with the central dogma arises from his technique for identifying sources of information. To give an example, in many organisms sex is

*See Barry Commoner "DNA and the Chemistry of Inheritance." *American Scientist* 52 (1964), 365-388.

determined in a superficially simple manner by chromosomes: individuals carrying a pair of so-called X-chromosomes are female; those carrying one X- and one Y-chromosome are males. Males produce two types of gametes, X-bearing and Y-bearing; offspring are of two sorts, males (XY) and females (XX). All eggs are alike as far as chromosomal content is concerned; according to Commoner, the *information* concerning the sex of the offspring resides in the sperm. An X-bearing sperm carries information for (or specifies) femaleness while a Y-bearing sperm carries information for (or specifies) maleness. In diagrammatic form, this account appears as follows:

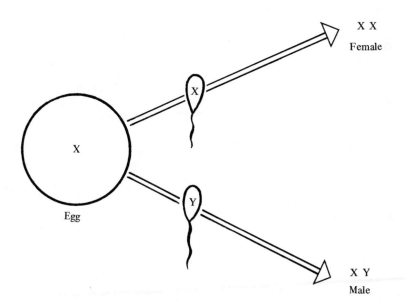

where ⊏════▷ can be read: "with information provided by." The diagram, consequently, can be reduced to prose as follows: An egg, with information provided by an X-bearing sperm, produces a female offspring; alternatively, an egg, with information provided by a Y-bearing sperm, produces a male offspring.

At this time it is necessary to place a limit upon the extent to which the diagram shown immediately above can be profitably used as an analytical tool.

We might, for example, draw what appears to be a similar diagram:

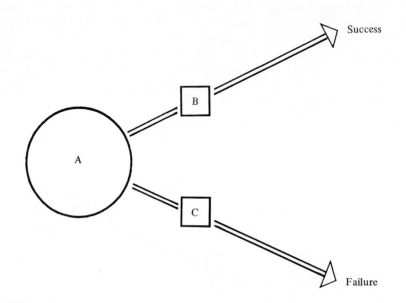

Given a dichotomy in which the alternative end points are *Success* and *Failure* (or *Yes* and *No*), it is tempting to ascribe the information or specificity to the alternatives *B* and *C* if *A* is given as a constant; superficially, this appears to be identical to the case with sperms and eggs illustrated earlier. Upon reflection, however, it appears that given a sufficiently complicated *A,* there is no set of alternatives *B* and *C* that cannot specify *Success* or *Failure.* Thus, whether I *walk* or *ride* to work can carry information concerning the success or failure of a sufficiently complicated biological experiment; one act can lead to success, the other to failure. When information can be shown to reside without exception in every conceivable object or any act which one might perform, information as it has been defined here ceases to be a useful concept.

In the case of dichotomous diagrams leading to two specific end points out of many possible ones (male and female out of many possible intersexes, supersexes, or inviable individuals), the notion that information is carried by the given alternatives is both valid and valuable. And here Commoner makes the following two points:

> (1) The enzyme (enzyme$_1$ in the following diagram) that actually assembles DNA from its precursers will, if primed, faithfully replicate a sample of DNA but, in the absence of primer DNA, the enzyme may assemble two different types of DNA. In the absence of primer, according to Commoner's reasoning, the specificity for DNA structure must reside in the enzyme (a protein).

(2) The enzyme (enzyme$_2$ in the following diagram) that prepares amino acids for assembly into polypeptide (=protein) chains, carries out its work so that a sequence of these nucleotide bases in the messenger-RNA specifies a given amino acid at a given site in the protein molecule. If this enzyme is extracted from one organism (say, a rabbit) and is forced to work in the cellular machinery of a second (a bacterium for example), it may consistently err by substituting one amino acid for another when attempting to insert the second in the growing protein molecule. Thus, Commoner claims, specificity for amino acid sequence does not lie entirely with the messenger-RNA; it lies in part with the enzyme.

Having shown that some of the specificity for DNA structure resides in the enzyme (a protein) responsible for DNA replication and having shown that some of the specificity for the amino acid sequence of a protein lies with another enzyme (another protein) that aids in the translation of messenger-RNA, Commoner obtains a diagram such as the following (where the open arrows are to be read "with information provided by" as before):

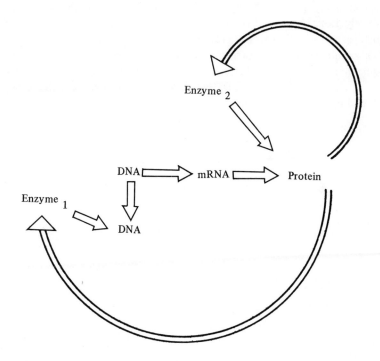

Instead of the linear diagram of the central dogma, Commoner, on the basis of his analysis, obtains a diagram consisting of at least two interconnected cycles in which enzymes (proteins) have an opportunity to specify the nature of the DNA, which in turn specifies protein structure. The evolution of DNA structure according to this new diagram could be under the control of proteins since specificity for the replication of DNA resides in part with the enzyme involved in replication.

The scheme devised by Commoner does not seem to apply to any one species. It is a scheme that works if (1) because of highly artificial experimental conditions, there is no DNA available so that enzyme$_1$ is forced to improvise and if (2) cellular machinery of widely different organisms are forced to work together in test tubes. Within a single species, where DNA is always present to act as its own primer and where the cellular machinery has been developed during the species' evolutionary history, the complications cited by Commoner do not exist. The central dogma of molecular biologists works in the linear fashion shown in the opening diagram; proteins do not specify the composition of mRNA, and mRNA does not — usually — specify the composition of DNA. Evolution, therefore can be looked upon as a process, acting through natural selection, by which the complications envisaged by Commoner (and demonstrated experimentally by others) are prevented from muddying the attractive simplicity of the central dogma. The origin of new species that are reproductively isolated from their near relatives may at times reflect the extent to which natural selection acts in preserving the essential truth of the central dogma while simultaneously permitting its details to evolve; the extent to which the dogma holds even in synthetic, interspecific mixtures of messenger-RNA and other metabolic machinery suggests, however, that the basic features of the central dogma evolve exceedingly slowly.

Genetic Engineering:
The Promise of Things to Come*

Radioactive isotopes trace the course of chemical elements through metabolic pathways; heavy isotopes and high-speed centrifuges separate substances that are otherwise chemically identical; and a battery of newly devised chemicals stop the machinery of living cells in the act of performing specified reactions. These are the research techniques of the molecular biologist. These same techniques have enabled geneticists to make advances in the past two decades that were undreamed of before World War II. The physical structure of DNA is known; the basis for erroneous replication of genetic material is understood at the physicochemical level; the genetic code has been broken so that the correspondence of amino acids in protein molecules to nucleotide sequences in DNA is understood; and even the punctuation marks of genetics – nucleotide sequences whose role it is to start or stop the synthesis of polypeptide chains – have been identified. Genes of known sorts can be introduced into bacterial cells by either bathing them in DNA solutions prepared from the proper material or infecting them with phage particles grown on other bacteria.

The ease with which geneticists manipulate genes (DNA) in microorganisms has encouraged them to look with near impatience at man and his inherited ills. At one time, to be told that a disorder was "genetic" was equivalent to being told it was "incurable." Now genetic disabilities are yielding to therapy devised through an understanding of the precise nature of the genetic defects involved. *Eugenics* and *genetic engineering* are terms that cover various corrective actions against genetic disabilities where the rationale of these actions is based on detailed knowledge of gene function rather than, for instance, gross features of an abnormal, hereditary syndrome.

Relief for the genetically handicapped can take several forms. A common feature of mutant genes is the failure of each to produce its corresponding enzyme in quantities sufficient for carrying on its normal function. The lack of an enzyme may lead to a deficiency of a substance essential for life or, if the enzyme is one that destroys a waste product, may allow the accumulation of a toxic chemical. Once the biochemical pathway has been identified, the affected

*Adapted in part from Bruce Wallace,"Genetics and Genetic Manipulations" in Arnold B. Grobman, ed., *Social Implications of Biological Education* (Washington, D.C.: National Association of Biology Teachers, 1970).

individual can often be treated by supplying him with the missing substance (insulin injections for diabetics are an example) or by withholding the offending one (phenylalanine can be omitted from the diet of children suffering from phenylketonuria; milk can be witheld from those children who lack lactase, a normal digestive enzyme). A treatment that consists of supplying a missing gene product (euphenics) does not repair the underlying genetic defect; it merely bypasses the trouble caused by the mutant gene. The affected person, should he have children, will transmit his defective gene to his descendants according to the standard rules of Mendelian inheritance.

A more sophisticated remedial procedure suggested by modern genetic techniques is one involving the activation or suppression of gene action in certain cells of the body. This as-yet hypothetical procedure would be effective for those diseases caused by the improper control of gene action rather than by defective gene products. Suppose, as an example, that hemoglobin, a protein whose synthesis is normally restricted to red blood cells, were in certain persons to be made by cells of other tissues such as those in the cornea of the eye. A disease of this sort would be characterized by poor vision and the presence of reddish-brown blotches in corneal tissue. Following regulatory therapy, the genes controlling hemoglobin production in the corneal cells would be rendered inactive so that the corneas would become transparent as they should be. Again, however, the treatment would not cure the underlying genetic defect; it would remove the abnormal characteristic but would leave the "cured" individual free to pass on his abnormal gene to his descendants.

A third procedure for curing genetic diseases (*genetic engineering* in the strict sense) involves the introduction of normal genes into the body cells of the affected individual by means of infectious viruses. A virus consists of a small segment of DNA surrounded by a protein coat. The coat is responsible for the virus's infectious properties. The DNA enters the cell and, in a typical case, takes control of the cell's genetic machinery and redirects it so that large quantities of virus DNA and coat protein are produced as end products. By wrapping virus coats around many artificially manufactured replicas of a normal human gene, scientists could infect a genetically afflicted person so that a large number of his body's cells now would possess the normal gene. If the gene product were one that circulated through the body, its production by some cells — even though not by all — should remove the characteristically abnormal symptoms.

Doctors and technicians who have worked with a particular rabbit tumor virus, the Shope papilloma virus, illustrate the type of effect that could be expected following "infection" by normal genes. One of the genes of the papilloma virus is responsible for the production of an easily recognized enzyme, arginase (an enzyme that breaks down the amino acid *arginine* into *ornithine* and *urea*). Workers who have handled this virus are frequently infected, a fact revealed by both the abnormally low concentrations of arginine in their blood and the presence of the atypical enzyme, one that is quite different from any normally found in man. Dr. Joshua Lederberg has pointed out that had there existed a human disease characterized by high levels of arginine in the blood, genetic engineers would be proclaiming their first medical triumph. As it is, the virus seems to be entirely nonpathogenic in man and the new enzyme seems to have no effect on him. The substitution of normal for mutant genes in body cells of a genetically defective person might cure his disease but, as in the case of the "cures" mentioned above, the genes he would pass on to his children would still be defective.

The last remedial procedure that is dreamed of by molecular biologists consists of the substitution within the germ cells themselves of normal genes for those known to be defective. Presumably this treatment would involve the bathing of germ cells with a specific DNA solution prior to their use in fertilizing an egg. If such treatment were successful, the passage of the mutant gene from one generation to the next would be halted.

The possibility for carrying out these, and still other, genetic tricks certainly exists. Whether they will in fact be carried out in the near future is problematical. Higher mammals, including man, are much more complicated than bacteria. The regulation of gene action is a complex problem in a higher organism and is one that in the past biochemists have consistently under-estimated. Not long ago we read that the active site of an enzyme is the only essential part of an enzyme molecule; the rest of the protein molecule according to this simple view was made by "fossil" DNA. More recently, allosteric properties of enzymes have been discovered; these involve a change in the physical configuration of enzyme molecules when they combine with certain small molecules that act as enzyme activators or suppressors. Allosteric changes represent a form of enzyme regulation; they call for precise and sophisticated molecular structures. In my opinion, problems of intercellular communication and the proper regulation of tissue and organ growth within higher animals multiply much more rapidly than does the obvious physical complexity of the animal itself. Thus, I am inclined to doubt whether some of the more exotic euphenic measures that have been suggested will be successful.

Quite apart from the matter of success, however, there are divergent views regarding the desirability of these engineered changes. Dr. Marshall W. Nirenberg has urged that we refrain from such tamperings until we can see precisely what we are doing. Dr. Joshua Lederberg has spoken out for experimentation whenever experimentation appears to be called for. Indeed, as Lederberg has pointed out, a physician might properly be reprimanded in a given instance for *not* attempting a promising new treatment if other remedial procedures have failed.

I tend to favor Lederberg's point of view. I am particularly impressed by his emphasis on the individualistic nature of the experimental stage – one patient, one doctor. To delay the early experimental trials while investing tremendous sums of money in the development of a potential remedial procedure causes pressures (face-saving, among other kinds) for its largescale use. And so, if an error in judgment has been made, it will have an enormous impact on the population; this is the outcome I fear most of all. To cite one example, a simian virus has been widely spread in the human population as a contaminant of early polio vaccinations; we are still waiting patiently to discover what effect, if any, it has on man. We simply do not know. As a matter of personal preference, I would rather risk small-scale successes and small-scale errors than large-scale successes and large-scale errors. Success can be amplified; a blunder, once committed, cannot be retracted.

A spector that haunts many minds is one in which genetic control of citizens is taken over by the government. Anything is possible, of course, but I prefer to think that citizens will not take leave of their senses en masse. If I did not believe this, there would be remarkably little in life that would cheer me. If I could not assume that the government represents people and that the majority of persons are sane and reasonable, I could become terrified at the thought of artificial insemination, lack of gun controls, compulsory military training, income taxes, and a host of other unpleasant aspects of communal life. Several of the items named do depress me, to be sure, but I have never really expected the government to take *all* of my income, or to make me serve my *entire* life in the army. By the same token, I do not expect the government to remodel me or my children genetically according to some mysterious, diabolical plan.

There remains, however, a problem that *is* raised by euphenic measures. By surviving where, in earlier times, they would have died, individuals carrying defective genes are enabled to pass on defective genes from generation to generation within the population. Should we worry about this? Should we permit it? The obvious answer is the algebraic one provided by population genetics. The frequency of a mutant gene in a population increases to a level commensurate with the deleterious effect it exerts on its carriers. Hence, if euphenic measures mask all deleterious effects of a gene, there should be no harm even if the frequency of the gene were to reach 100 per cent. This line of reasoning assumes, however, that populations approach new equilibrium conditions smoothly. Recently I have been impressed by the irregularity of population changes. The potato famine in Ireland during the 1840's represents either a zig or a zag on the curve depicting Ireland's approach to an equilibrium population; 2½ million of 9 million persons either starved or emigrated during this "irregularity." Catastrophes of this sort should be avoided. Unforeseen events might easily interfere with an otherwise standard euphenic treatment of a large number of persons and thus lead to a large-scale catastrophe for the population to which they belong. To avoid this possibility, I would prefer that persons saved by euphenic measures be discouraged from bearing children. This request in the very near future need not be a traumatic one because some sort of population control will be needed at any rate and, under these control measures, many persons will remain childless. The abstention from reproduction on the part of each person who has been aided by euphenic measures would, on the average, avert a human tragedy in some future generation.

The Impact of
Genetic Disorders on Society

In a small, best-selling paperback volume entitled *Heredity, Race, and Society*, Professsors L. C. Dunn and Th. Dobzhansky describe a conflict between "tasters" and "nontasters" over water purification procedures in an imaginary town. Tasters are persons to whom a certain chemical (phenylthiocarbamide, or PTC) tastes extremely bitter; nontasters (of whom I am one) are unable to detect any taste to this chemical. Dunn and Dobzhansky describe the reaction of townspeople to the use of PTC as a disinfectant in the town's water system. The tasters (some 30 per cent of the population) are violently upset by the practice, but the nontasters (the remaining 70 per cent) are highly pleased. An inspector who is sent from the state capital is, by chance, a nontaster; in his report to the governor he refers to tasters as prejudiced troublemakers who arbitrarily oppose technological progress. Finally, a local referendum is held but the tasters are voted down by the majority who are nontasters. In the eyes of the minority this act amounts to tyranny by the majority, and it appears that rebellion is the only recourse left to the tasters.

The hypothetical example cited by Dunn and Dobzhansky is an extremely simplistic one; for example, members of a single family may differ in their tasting ability and so, in visualizing the coming rebellion, we are left with the unhappy thought of divided families whose members — mothers, fathers, brothers, and sisters — are at each other's throat. At least one political scientist was unable to continue beyond this early section of *Heredity, Race, and Society*. If this example illustrates a geneticist's view of the origins of rebellion in human society, he doubted that he would find anything of value elsewhere in the book.

In one sense the political scientist was right, of course; social upheavals have complex causes that trace back to many different sources. No matter how simply the two sides are designated once the fighting begins, the underlying causes are never as simple as tasting and nontasting. On the other hand, genetic variation is not always obvious even when the consequences of this variation are far from trivial; in the shaping of a society under these circumstances, single genes may play a role that is surprisingly large. In this essay I shall describe two cases of simple genetic disorders having important impacts on society. The first case concerns the revolutionary origins of the United States; the second case concerns the health of the black community of the inner city — a segment of our society that has become restive for many different reasons.

The American colonies waged the Revolutionary War and won their independence from Britain during the reign of King George III, a ruler commonly regarded as mad. The *Columbia Encyclopedia* refers to his "intermittent periods of insanity." The consensus on both sides of the Atlantic Ocean is that he must have been insane to sign the Stamp Act, the act that evoked the colonists' slogan of "no taxation without representation" and led eventually to the outbreak of hostilities.

A recent reexamination of notes and records of George III's attending physicians has revealed that the king's illess was genetic in origin. He suffered from porphyria, a disease caused by the body's inability to properly metabolize discarded hemoglobin molecules. The attending doctors noted on several occasions that the king's urine was red or discolored; this symptom arises from the excretion of porphyrins, pigments that are normally converted by the body's cells into smaller molecules.

Porphyria is characterized by a number of symptoms: sensitivity to light, colic, nausea, weakness, visual disturbances, headaches, tremors, and convulsions. All of these are reflected in notes that refer to King George's delirium and hallucinations. Furthermore, upon examination of the records that contain the medical histories of blood relatives of the king, it appears that the disease affected Mary, Queen of Scots (in the late 1500's), King James, George IV, and others of three interrelated royal houses. A number of present-day descendants of George III have a form of porphyria that causes their skin to be especially sensitive to sun and injury.

In this example, we see a genetic disorder that would have responded in part to a proper diet and rational treatment (even in the 18th century). Instead, untreated in a single individual, it may have influenced the course of history for English-speaking peoples.

The second genetic trait, one that is today probably affecting the health of hundreds of thousands of inner-city residents, is sickle cell disease. This is a disorder that affects Negroes almost exclusively. The name of the disease comes from the misshapen appearance of red blood cells in solutions from which oxygen is excluded. Sickling disease is of two sorts. Persons suffering from the milder *sickling trait* carry one mutant and one normal gene of a pair that is involved in hemoglobin synthesis; those suffering from the much more severe (generally fatal) *sickle cell anemia* carry two mutant (and, hence, no normal) genes of this sort.

The prevalence of the sickling gene in Africa seems to be related to the high incidence of malaria in certain regions of that continent; other hemoglobin defects are common in human populations inhabiting other malarial regions of the world. Persons carrying one normal and one mutant (sickling) gene are resistant to some forms of malaria whereas individuals with normal red blood cells are susceptible to malaria, and in regions where the disease is endemic, a great many of these "normal" persons die of it. Thus, in the malaria-infested regions of Africa the surviving persons are (to exaggerate somewhat) those carrying one normal and one mutant gene at the sickling locus. The children of these persons die either of anemia (one-fourth of all children born) or of malaria (again, roughly, one-fourth of all children born) or, in the case of those children with one gene of each sort, they survive (one-half of all children born).

An individual who carries only one sickling gene exhibits no gross, outwardly visible clinical symptoms. Under laboratory tests his red blood cells will assume the diagnostic sickle shape, to be sure; nevertheless, his daily activities are normal and he appears to be as resistant to infectious diseases as other persons. He does suffer at high altitudes, however, because his red blood cells contain only 60 per cent to 80 per cent as much functional hemoglobin as do the red blood cells of nonsickling individuals.* The remainder of the hemoglobin is abnormal (in the mutant hemoglobin one amino acid molecule in a chain of 146 has been replaced by another) and is unable to transport oxygen. In a sense, a person who carries a sickling gene behaves at sea level much as a normal person would at high altitude; his body compensates for the decreased oxygen-carrying capacity of the red blood cells by increasing the number of them in circulation and by other physiological adaptations.

The sickle-cell trait offers its carriers no advantage in a geographic region where malaria is rare or nonexistent. Consequently, American Negroes have substantially lower frequencies of the responsible gene than do their putative ancestors. At the present time, the frequency of persons with the sickling trait is about one in ten within the black American population. We shall now abandon the genetics of hemoglobin in order to discuss quite another problem: carbon monoxide poisoning.

Carbon monoxide (CO) is formed during the incomplete combustion of any carbonaceous substance. It is an extremely poisonous gas because it combines with hemoglobin three hundred times as readily as does oxygen. A small concentration of CO, therefore, can effectively keep the blood from picking up oxygen in the lungs and transporting it to the tissues of the body. Continued exposure to concentrations of CO of some 1,500 to 2,000 ppm (parts per million) results in the loss of consciousness and, in a short time, death.

Persons are exposed to various amounts of carbon monoxide throughout each day. Inhaled cigarette smoke contains about 400 ppm of the gas. Automobile exhaust fumes contain high concentrations of carbon monoxide; over 8 million pounds are discharged by street traffic in New York City each

*On February 7, 1970, the *New York Times* reported that the deaths of four Negro recruits who were performing calisthenics at high (4,000 feet) altitude at Fort Bliss (El Paso), Texas, were caused by sickle-cell trait. A number of other deaths of Negro recruits at other Army installations are suspected of being caused under similar circumstances by the same trait. An Army medical officer has now urged that all Negro recruits be screened for this trait — a call that follows by two decades or more knowledge within the military that blacks frequently carry the responsible gene!

day, 20 million pounds in Los Angles. The level of carbon monoxide in large cities commonly exceeds 50 ppm; in areas of local traffic congestion the level may reach 150 ppm or more. A third source of exposure to carbon monoxide (in this account I am omitting occupational exposure, an exposure that may, for example, be very high in the case of a garage mechanic or traffic patrolman) is from within the body. The metabolic breakdown of hemoglobin leads to the endogenous production of carbon monoxide that then attaches to and inactivates a portion of the hemoglobin of normal red blood cells.

The point I want to make about the black communities of metropolitan ghettoes now becomes clear. An average of one Negro in ten carries the mutant gene responsible for the sickling trait. The red blood cells of these persons contain only 60 per cent to 80 per cent functional hemoglobin (although physiological adaptation may alleviate this deficit somewhat). Very often the ghettos of the inner city border on or straddle the main traffic arteries linking the city to nearby suburbs and airports. The level of carbon monoxide from auto exhaust fumes near these arterial highways is high at all times but is virtually unbearable during morning and evening rush hours. The ghetto inhabitant is quite likely to be a moderate or heavy smoker; for many of these persons the cigarette habit begins during childhood. The emerging picture shows that 10 per cent of the black, inner-city population must be suffering from severe chronic carbon monoxide poisoning. The long-range effects of such poisoning are known to include tissue damage both in the central nervous system and in the heart muscle. High levels of carbon monoxide are also known to interfere with reaction times and to impair judgement. Here, therefore, the cost of technology has fallen heavily upon one segment of the population.

The case of children born in the ghettos to black women who suffer from sickle cell disease is especially serious, it seems to me. During pregnancy the developing fetus places tremendous physiological demands on its mother and, indeed, needs to have these demands met if its own prenatal development is to proceed normally. In addition to the maternally restricted prenatal environment, half of the children born of sickling mothers are themselves carriers of the sickling gene. These persons, in addition to their prenatal handicap, suffer a chronic oxygen deprivation during their early childhood years when much of the physical machinery of their bodies is supposedly being perfected. The question was once put to me, "Must the environment be tailored to meet the needs of the most sensitive among us?" I believe the instinctive answer to this question must be, "Yes." I see no reason to equivocate in the vain hope of dredging up a convincing rationalization.

Genetics Under Stalin:
The Suppression of a Science

From July 31 until August 7, 1948, the Lenin Academy of Agricultural Sciences held an extraordinary meeting. Its minutes are available in book form.*

The President of the Academy was Trofim D. Lysenko. Because an address by Lysenko was the main item on the agenda, the meeting was chaired by Academician P.P. Lobanov. The subject of the meeting was an inquiry into the status of genetics and the beliefs and research activities of geneticists. Lysenko's address includes numerous passages like the following:

> To us it is perfectly clear that the foundation principles of Mendelism-Morganism are false. They do not reflect the reality of living nature and are an example of metaphysics and idealism.
>
> Because this is so obvious, the Mendelist-Morganists of the Soviet Union, though actually fully sharing the principles of Mendelism-Morganism, often conceal them shamefacedly, veil them, conceal their metaphysics and idealism in a verbal shell. They do this because of their fear of being ridiculed by Soviet readers and audiences. . . .

● ● ●

Here is one example which might be cited to show how useless is the practical and theoretical program of our domestic cytogeneticists.

Professor of Genetics, N. P. Dubinin, Corresponding Member of the Academy of Sciences of the U. S. S. R., who is regarded by our Morganists as the most eminent among them, has worked for many years to ascertain the differences in the cell nuclei of fruit flies in urban and rural localities.

His work is entitled "Structural Variability of Chromosomes in Populations of Urban and Rural Localities.". . .

But if we were to describe his work in plainer terms, stripping it of the pseudoscientific verbiage and replacing the Morganist jargon with ordinary Russian words, we would arrive at the following:

As the result of many years of effort Dubinin "enriched" science with the "discovery" that during the war there occurred among the fruit-fly population of the city of Voronezh and its environs an increase in the percentage of flies with certain chromosome structures and a decrease in the percentage of flies with other chromosome structures. . . .

Dubinin sets himself further tasks for the restoration period. He writes:

'It will be very interesting to study in the course of several coming years the

*Lenin Academy of Agricultural Sciences of the U.S.S.R. 1949. *The Situation In Biological Science*; Proceedings 1948. International Publishers, New York.

restoration of the karyotypical structure of the urban population in connection with the restoration of normal conditions of life.' (Animation. Laughter.)

That is typical of the Morganist's "contribution" to science and practical activity before the war and during the war, and those are the vistas of the Morganist "science" for the restoration period. (Applause.)

Finally, a nearly unbelievable passage reveals the true authority upon which the entire proceedings of the Academy were based:

> Comrades, before I pass to my concluding remarks I consider it my duty to make the following statement.
>
> The question is asked in one of the notes handed to me, What is the attitude of the Central Committee of the Party to my report? I answer: The Central Committee of the Party examined my report and approved it. (Stormy applause. Ovation. All rise.)

It is now clear, as Z. A. Medvedev has reported,* that Lysenko had asked Stalin to sanction the proposed rout of geneticists. The original text of Lysenko's report, with Stalin's personal corrections, was kept in Lysenko's office.

This meeting marked the end of formal genetics in Soviet schools and universities for nearly two decades. Its closing sessions consisted largely of public recantations by many geneticists who prudently acknowledged the great wisdom of their leader, Joseph Stalin. Few Russian biologists during the following years were able to continue their work unmolested; by the authority of governmental decrees and by means of the most unabashed witch hunts, genetic concepts and genetic terminology were ruthlessly expunged from all biological sciences. The giants of Russian genetics were affected most of all; they were virtually enemies of the State. Molecular genetics was rescued from total oblivion in the Soviet Union by atomic physicists; molecular biologists held positions in physics institutes because of the important but poorly understood biological (chiefly genetic) effects of radiation. Consequently, under the protection of prestigious nuclear physicists, a vestige of modern genetics was saved.

What is the background for these amazing events? Who is Trofim Lysenko? How did he achieve such power? What is the status of Lysenkoism today? These are the varied topics of this essay.

Genetics has always occupied a peculiar position in biology. The logic of geneticists and their method of experimentation have not always been appreciated by their colleagues in other branches of biology. That Mendel's

---

*Z. A. Medvedev, *The Rise and Fall of T.D. Lysenko* (New York: Columbia University Press, 1969). Medvedev is a Russian biologist and a member of the Soviet Academy of Sciences. During the summer of 1970, the Russian government attempted to commit Medvedev to an insane asylum because of alleged schizophrenia. (He insists on expressing views on topics outside his scientific specialty.) Only a public outcry by his fellow academicians prevented his summary incarceration.

research lay forgotten for 35 years is evidence for this claim. At the time Mendel's results were rediscovered, physiologists and experimental embryologists occupied the center of the biological stage. The great importance attached to what are now called "Mendel's Laws" was placed there by his discoverers. Mendel himself had used his observed ratios to reach a decision about the pattern of inheritance, and his later work, at least in part, was designed to reveal the physical nature of the hypothetical entities he had postulated on purely theoretical grounds. Professor T. H. Morgan, while he was still an embryologist rather than a geneticist, claimed that if a geneticist cannot explain his observations by one gene, he will invent two. The quotation from H. J. Muller's writings that has been used to introduce another essay in this section also reveals the strain that existed between early geneticists and their contemporary physiologists.* Even today there are biochemists and ecologists, members of two rather separate biological sciences, who marvel that molecular geneticists and ecological geneticists prefer to discuss research problems among themselves rather than to dismember genetics into what appear to be logically separable disciplines. These persons overlook the basic intellectual community that *is* genetics and which attracts geneticists.

Physiology has been a strong science in Russia for decades. Pavlov and his studies on the conditioned reflexes in dogs gave the biological sciences in Russia an impetus and a *direction* that can still be detected seven decades later. Behavioral sciences are of special interest to Communists because, they claim, it is under Communism that true equality among persons is to be attained. No individual is to be slighted; through training and education and with the benefits accrued under a benevolent Communist society, heaven is to be created here on earth.

Early geneticists, largely because they had no notion of the mechanisms by which genes exert their influence during development, were not nearly so convinced about the malleability of individuals. To say that a certain disease was genetic in origin meant, for all practical purposes, that it was incurable. The influence of the environment was dutifully acknowledged but, in the absence of information concerning gene action or, more important, the specific action of particular genes, environmental modification appeared to be an ineffectual means for offsetting genetic ills. To a great many of these early geneticists, particularly to those who flocked to eugenic movements, populations were to be improved by artificial selection through culling – this was true for populations of grains, of farm animals, of race horses, and of people.

Among the first signs that a seriously uneasy relationship existed between the government and the geneticists of the Soviet Union occurred with the cancellation of the Seventh International Congress of Genetics that was to be held in Moscow during 1937. Another was the failure of Russian geneticists to

*See the essay entitled "The Control of Physiological Processes."

attend the Seventh Congress that was finally held in Edinburgh, Scotland, two years later in 1939. Events built up with incredible speed and culminated in the arrest and imprisonment in August, 1940, of Academician N. I. Vavilov, one of the world's most renowned plant breeders and at that time President of the Soviet Academy of Sciences. Vavilov died in prison two years after his arrest.

Trofim Lysenko was a plant physiologist by training, apparently a man with rather shrewd insights into the needs of and care for farm crops. Having no understanding of genetics as either a theoretical or applied science, Lysenko interpreted all phenotypic alterations, whatever their source, as potentially heritable alterations. Their mere induction by physiological means was sufficient, in his opinion, to render these changes heritable, to insure their passage from one generation to the next. The idea that the germ plasm is relatively immutable and the notion that genetic material contained in a number of tiny chromosomes controls development were foreign to his concept of life. Lysenko had observations that he thought were fatal to the geneticist's point of view (for example, the conversion of winter to spring wheat by storing germinated grain in the cold). Such observations, made by others nearly a century earlier, have no bearing on genetic theory. Although the contamination of seed supplies best accounts for the supposed transformation of wheat into rye, rye into wheat, barley into oats, and the many other transformations reported by Lysenko's followers, Lysenko grasped at these bizarre data as support for his antigenetic views rather than as observations that needed careful repetition.

A lawyer and political theorist, I. I. Prezent, joined Lysenko during the early 1930's; together they promoted Lysenko's views as ones consistent with Marxism whereas those of geneticists were said to be contrary to Communist doctrine. With the out-and-out entry of politics into the scientific arena (especially during Stalin's era), we can imagine the ensuing course of events. First, there was the allotment of state funds for research: certain budgets were cut at the expense of others; certain research projects were left entirely unsupported; and only "proper" research papers were printed in the scientific journals. Second, we can imagine the intellectual accommodation that many scientists made (scientists are people too) in order that their research would receive financial support, a compromise here and a change in emphasis there are to many scientists seemingly tolerable accommodations if such are needed in order to continue research. Third, because the allotment of funds requires administrative personnel, we can also imagine the superstructure of clerks and administrators who, in response to political arguments, diverted the flow of money and facilities in favor of Lysenko's followers. Finally, the stage was set for an outright confrontation between Lysenkoists and Mendelist-Morganists, of which excerpts have already been quoted.

Entanglements involving science, politics, and money are not undone overnight. Stalin passed away, and it appeared that Lysenko would become an obscure figure on Russia's scientific scene, but he still clung to his high scientific positions. At one time a government commission seemed ready to censure him but, to the surprise of many, Khrushchev turned to Lysenko in his efforts to improve Soviet agriculture. He said, in effect, that Lysenko's views "made sense"; genetic concepts, I might emphasize, are not intuitively obvious. Khrushchev's successors finally destroyed Lysenko's reign; he has by now lost all of his prestigious posts.

Old-time Russian geneticists have been dusted off and placed once more in their original positions. Vavilov, together with numerous colleagues, died during imprisonment, but their names have now been refurbished. Many geneticists, however, despite official refurbishing, have lost the productive years of their lives without an opportunity for training young students. Genetics has improved in the Soviet Union, thanks in part to the biologists who were sheltered in the physics institutes, but even now an occasional genetics publication in a Russian journal exudes a Rip Van Winkle aura.

Are there lessons for others in the bizarre details of the Lysenko affair? The first emerges from Lysenko's rude remarks to Dubinin — ridicule aimed at Dubinin's irrelevance in wartime and postwar Russia. There are cries today for relevance; many have a solid basis. Pure research, the expansion of knowledge for knowledge's sake alone, is a luxury only the wealthy nations can readily afford. University scientists in most developing nations are expected, and legitimately so, to work on "practical" problems as well as their favorite theoretical ones. By analogous reasoning, many young scientists and engineers in the United States today have decided that at least part of their professional efforts should be spent on the social aspects of technology; in their estimation the United States can no longer neglect the unwanted and dangerous side effects of an advanced technology.

The accusation of irrelevance made by those who stand to gain as a consequence is a dangerous maneuver. Lysenko used the accusation of irrelevance against his enemies in an effort to gain further advantage over them in what was basically a struggle for political power. Today there are those who would stop all scientific research because, as they explain, knowledge is dangerous. When this plea is made from conviction, it is misguided; when it is made as a ruse to shift financial support from the sciences, it is evil. There is no more danger in uncovering hitherto unsuspected natural phenomena or relations between these phenomena than in uncovering hitherto unknown historical or archaeological facts. Both exist whether we are aware of them or not; both can be used for undesirable purposes by those who are intent upon doing so.

When public funds are the source of much scientific research, the task of keeping politics out of science becomes an impossible one. Politics must enter into decisions whenever limited funds have to be allocated among competing projects. The dangers that might otherwise accompany the intrusion of politics into research are lessened by keeping the public well-informed. Occasionally a member of Congress attempts, much as Lysenko did, to make political hay by ridiculing government-sponsored research projects; this infantile practice has lost its effectiveness in recent years. More insidious are the professional publicity campaigns waged by the promoters of large research projects in order to sell these to the public and, through the public, to Congress. In these cases the claims – made in effect by advertising agencies – must in every instance be subjected to the sharpest possible intellectual criticism. Prezent was Lysenko's publicity agent; he was the one who sold Lysenko's naive views to receptive governmental officials. Publicity men who sell research projects are interested first in the sale and only secondarily, if at all, in the quality of their merchandise. These publicity men are the Prezents of American science.

SECTION TWO

Evolution

Introduction

Darwin was a superb naturalist. His publications ramble leisurely from one fascinating biological problem to another, for example, a comparison of cross- and self-fertilization in plants, the origin of coral reefs, sexual selection, the descent of man, the role of earthworms in the formation of vegetable mold, the movement of climbing plants, and the contrivances by which orchids arrange their fertilization by insects. *Origin of Species* is, of course, a classic in biology, a monumental work whose thesis after more than a century remains essentially unchanged despite the rather poor state of peripheral information upon which Darwin leaned in its preparation.

The impact of Darwin's book on modern thought is beyond measure. In *The Making of the Modern Mind,* a marvelous old textbook that was the bugbear of an earlier generation of Columbia undergraduates, J. H. Randall, Jr. lists six of the more profound influences *Origin of Species* has had on man:*

First, the emphasis it placed on detailed causal analyses of specific processes of change; the rejection of a static truth and the adoption of the investigation of numerous little truths.

Second, the replacement of physics and mathematics by biological attitudes and methodologies especially in the social sciences; the development of the notion that social groups themselves are one of man's adaptations.

Third, the removal of man as the center of attention and its replacement by *difference* – that is, differences between men and between groups of men.

Fourth, the introduction of a whole new set of values; the raising of "modern" and "up-to-date" to places of prestige and the downgrading of "rational" or "natural." [Here Randall cannot refrain from commenting that it is perhaps an open question if in our new scale of values we have not lost as much as we have gained.]

Fifth, the emphasis that human beings must play a role in social change if change is to eventuate in anything worthwhile. Progress is not inevitable; on the contrary, it requires guidance. With the growth of molecular biology, this notion has been extended to include man's control over his own *biological* evolution.

Sixth, the reinforcement of irrationalism. This last is a striking point, for if beliefs themselves are but evolutionary adaptations, truth is also but a particular adaptation. "Are not all our beliefs but more or less concealed rationalizations, the reasons we invent for believing what we really believe because of quite different influence?" And does this question not identify Darwin as the stage manager for the Theater of the Absurd?

*The following material is adapted from John Herman Randall, Jr., *The Making of the Modern Mind,* Rev. ed. (Boston: Houghton Mifflin Co., 1940), pp. 489-494.

The concept of evolution through natural selection has been used by men to justify the suppression of other races; it has been used by the wealthy to justify the exploitation of the poor; it has been used by strong nations to justify the subjugation of the weak. It has removed fixed points of reference from our lives and leaves us floating – to the terror of some – in a sea of perpetual change. Nevertheless, upon looking about many of us say with Darwin, "There is a grandeur in this view of life."

The selection chosen to introduce this section is the account of the Galápagos archipelago given by Charles Darwin in *The Voyage of the Beagle.* The plants and animals of these islands were responsible for converting Darwin to the notion that natural life of today has evolved from earlier forms of yesterday; the bits and pieces of evidence that were available to him and his thoughts just prior to his conversion are found in this selection.

The captain of the *Beagle,* Robert FitzRoy, was not converted by the evidence which Darwin painstakingly assembled during the ship's five-year voyage nor by Darwin's incessant intellectual probings and discussions. Twenty-five years after the return of the *Beagle* to England, the clergy of the Church and the members of the Royal Society confronted one another in debate at Oxford. It was here that Thomas Huxley insulted Bishop Wilberforce by informing him that he would prefer to be descended from an ape than from a man like the Bishop who prostituted the gifts of culture and eloquence to the service of prejudice and falsehood.

> And now something intensely interesting intervened. Amid the hubbub, a slight gray-haired man got to his feet. His thin, aristocratic face was clouded with rage, and he waved a Bible aloft like an avenging prophet. Here was the truth, he cried, here and nowhere else! Long ago he had warned Darwin about his dangerous thoughts. Had he but known then that he was carrying in his ship such a . . . .*

An intervening century has failed to wholly dissipate the views held by the captain of the *Beagle.* A montage of recent newspaper headlines follows the selection from *The Voyage of the Beagle;* this collection illustrates the emotion with which evolution is greeted even today.

*Alan Moorehead, "Annals of Discovery," *The New Yorker,* September 6, 1969, page 94.

The Voyage
of the Beagle

Charles Darwin

Galapagos Archipelago

SEPTEMBER 15th. — This archipelago consists of ten principal islands, of which five exceed the others in size. They are situated under the Equator, and between five and six hundred miles westward of the coast of America. They are all formed of volcanic rocks; a few fragments of granite curiously glazed and altered by the heat, can hardly be considered as an exception. Some of the craters, surmounting the larger islands, are of immense size, and they rise to a height of between three and four thousand feet. Their flanks are studded by innumerable smaller orifices. I scarcely hesitate to affirm, that there must be in the whole archipelago at least two thousand craters. These consist either of lava or scoriae, or of finely-stratified, sandstone-like tuff. Most of the latter are beautifully symmetrical; they owe their origin to eruptions of volcanic mud without any lava: it is a remarkable circumstance that every one of the twenty-eight tuff-craters which were examined, had their southern sides either much lower than the other sides, or quite broken down and removed. As all these craters apparently have been formed when standing in the sea, and as the waves from the trade wind and the swell from the open Pacific here unite their forces on the southern coasts of all the islands, this singular uniformity in the broken state of the craters, composed of the soft and yielding tuff, is easily explained.

Considering that these islands are placed directly under the equator, the climate is far from being excessively hot; this seems chiefly caused by the singularly low temperature of the the surrounding water, brought here by the great southern Polar current. Excepting during one short season, very little rain falls, and even then it is irregular; but the clouds generally hang low. Hence, whilst the lower parts of the islands are very sterile, the upper parts, at a height of a thousand feet and upwards, possess a damp climate and a tolerably luxuriant vegetation. This is especially the case on the windward sides of the islands, which first receive and condense the moisture from the atmosphere.

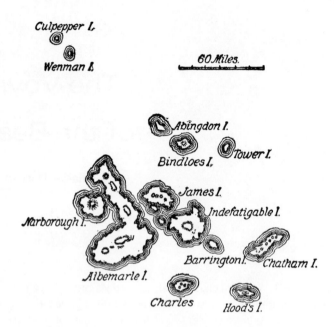

In the morning (17th) we landed on Chatham Island, which, like the others, rises with a tame and rounded outline, broken here and there by scattered hillocks, the remains of former craters. Nothing could be less inviting than the first appearance. A broken field of black basaltic lava, thrown into the most rugged waves, and crossed by great fissures, is everywhere covered by stunted, sun-burnt brushwood, which shows little signs of life. The dry and parched surface, being heated by the noon-day sun, gave to the air a close and sultry feeling, like that from a stove: we fancied even that the bushes smelt unpleasantly. Although I diligently tried to collect as many plants as possible, I succeeded in getting very few; and such wretched-looking little weeds would have better become an arctic than an equatorial Flora. The brushwood appears, from a short distance, as leafless as our trees during winter; and it was some time before I discovered that not only almost every plant was now in full leaf, but that the greater number were in flower. The commonest bush is one of the Euphorbiaceae: an acacia and a great odd-looking cactus are the only trees which afford any shade. After the season of heavy rains, the islands are said to appear for a short time partially green. The volcanic island of Fernando Noronha, placed in many respects under nearly similar conditions, is the only other country where I have seen a vegetation at all like this of the Galapagos Islands.

The Beagle sailed round Chatham Island, and anchored in several bays. One night I slept on shore on a part of the island, where black truncated cones were extraordinarily numerous: from one small eminence I counted sixty of them, all surmounted by craters more or less perfect. The greater number consisted merely of a ring of red scoriae or slags, cemented together: and their height above the plain of lava was not more than from fifty to a hundred feet; none had been very lately active. The entire surface of this part of the island seems to have been permeated, like a sieve, by the subterranean vapours: here and there the lava, whilst soft, has been blown into great bubbles; and in other parts, the tops of caverns similarly formed have fallen in, leaving circular pits with steep sides. From the regular form of the many craters, they gave to the country an artificial appearance, which vividly reminded me of those parts of Staffordshire, where the great iron-foundries are most numerous. The day was glowing hot, and the scrambling over the rough surface and through the intricate thickets, was very fatiguing; but I was well repaid by the strange Cyclopean scene. As I was walking along I met two large tortoises, each of which must have weighed at least two hundred pounds: one was eating a piece of cactus, and as I approached, it stared at me and slowly walked away; the other gave a deep hiss, and drew in its head. These huge reptiles, surrounded by the black lava, the leafless shrubs, and large cacti, seemed to my fancy like some antediluvian animals. The few dull-coloured birds cared no more for me than they did for the great tortoises.

23rd. — The Beagle proceeded to Charles Island. This archipelago has long been frequented, first by the bucaniers, and latterly by whalers, but it is only within the last six years, that a small colony has been established here. The inhabitants are between two and three hundred in number; they are nearly all people of colour, who have been banished for political crimes from the Republic of the Equator, of which Quito is the capital. The settlement is placed about four and a half miles inland, and at a height probably of a thousand feet. In the first part of the road we passed through leafless thickets, as in Chatham Island. Higher up, the woods gradually became greener; and as soon as we crossed the ridge of the island, we were cooled by a fine southerly breeze, and our sight refreshed by a green and thriving vegetation. In this upper region coarse grasses and ferns abound; but there are no tree-ferns: I saw nowhere any member of the palm family, which is the more singular, as 360 miles northward, Cocos Island takes its name from the number of cocoa-nuts. The houses are irregularly scattered over a flat space of ground, which is cultivated with sweet potatoes and bananas. It will not easily be imagined how pleasant the sight of black mud was to us, after having been so long accustomed to the parched soil of Peru and northern Chile. The inhabitants, although complaining of poverty, obtain, without much trouble, the means of subsistence. In the woods there are many wild pigs and goats; but the staple article of animal food is supplied by the tortoises. Their numbers have of course been greatly reduced in this island, but the people yet count on two days' hunting giving them food for the rest of the week. It is said that formerly single vessels have taken away as many as seven hundred, and that the ship's company of a frigate some years since brought down in one day two hundred tortoises to the beach.

September 29th. — We doubled the south-west extremity of Albemarle Island, and the next day were nearly becalmed between it and Narborough Island. Both are covered with immense deluges of black naked lava, which have flowed either over the rims of the great caldrons, like pitch over the rim of a pot in which it has been boiled, or have burst forth from smaller orifices on the flanks; in their descent they have spread over miles of the sea-coast. On both of these islands, eruptions are known to have taken place; and in Albermarle, we saw a small jet of smoke curling from the summit of one of the great craters. In the evening we anchored in Bank's Cove, in Albemarle Island. The next morning I went out walking. To the south of the broken tuff-crater, in which the Beagle was anchored, there was another beautifully symmetrical one of an elliptic form; its longer axis was a little less than a mile, and its depth about 500 feet. At its bottom there was a shallow lake, in the middle of which a tiny crater formed an islet. The day was overpoweringly hot, and the lake looked clear and blue: I hurried down the cindery slope, and, choked with dust, eagerly tasted the water — but, to my sorrow, I found it salt as brine.

The rocks on the coast abounded with great black lizards, between three and four feet long; and on the hills, an ugly yellowish-brown species was equally common. We saw many of this latter kind, some clumsily running out of the way, and other shuffling into their burrows. I shall presently describe in more detail the habits of both these reptiles. The whole of this northern part of Albemarle Island is miserably sterile.

October 8th. — We arrived at James Island: this island, as well as Charles Island, were long since thus named after our kings of the Stuart line. Mr. Bynoe, myself, and our servants were left here for a week, with provisions and a tent, whilst the Beagle went for water. We found here a party of Spaniards, who had been sent from Charles Island to dry fish, and to salt tortoise-meat. About six miles inland, and at the height of nearly 2000 feet, a hovel had been built in which two men lived, who were employed in catching tortoises, whilst the others were fishing on the coast. I paid this party two visits, and slept there one night. As in the other islands, the lower region was covered by nearly leafless bushes, but the trees were here of a larger growth than elsewhere, several being two feet and some even two feet nine inches in diameter. The upper region being kept damp by the clouds, supports a green and flourishing vegetation. So damp was the ground, that there were large beds of a coarse cyperus, in which great numbers of a very small water-rail lived and bred. While staying in this upper region, we lived entirely upon tortoise-meat: the breast plate roasted (as the Gauchos do carne con cuero), with the flesh on it, is very good; and the young tortoises make excellent soup; but otherwise the meat to my taste is indifferent.

One day we accompanied a party of the Spaniards in their whale-boat to a salina, or lake from which salt is procured. After landing, we had a very rough walk over a rugged field of recent lava, which has almost surrounded a tuff-crater, at the bottom of which the salt-lake lies. The water is only three or four inches deep, and rests on a layer of beautifully crystallized, white salt. The lake is quite circular, and is fringed with a border of bright green succulent plants; the almost precipitous walls of the crater are clothed with wood, so that the scene was altogether both picturesque and curious. A few years since, the sailors belonging to a sealing-vessel murdered their captain in this quiet spot; and we saw his skull lying among the bushes.

During the greater part of our stay of a week, the sky was cloudless, and if the trade-wind failed for an hour, the heat became very oppressive. On two days, the thermometer within the tent stood for some hours at 93°; but in the open air, in the wind and sun, at only 85°. The sand was extremely hot; the thermometer placed in some of a brown colour immediately rose to 137°, and how much above that it would have risen, I do not know, for it was not graduated any higher. The black sand felt much hotter, so that even in thick boots it was quite disagreeable to walk over it.

The natural history of these islands is eminently curious and well deserves attention. Most of the organic productions are aboriginal creations, found nowhere else; there is even a difference between the inhabitants of the different islands; yet all show a marked relationship with those of America, though separated from that continent by an open space of ocean, between 500 and 600 miles in width. The archipelago is a little world within itself, or rather a satellite attached to America, whence it has derived a few stray colonists, and has received the general character of its indigenous productions. Considering the small size of the islands, we feel the more astonished at the number of their aboriginal beings and at their confined range. Seeing every height crowned with its crater, and the boundaries of most of the lava-streams still distinct, we are led to believe that within a period geologically recent the unbroken ocean was here spread out. Hence, both in space and time, we seem to be brought somewhat near to that great fact — that mystery of mysteries — the first appearance of new beings on this earth.

Of terrestrial mammals, there is only one which must be considered as indigenous, namely, a mouse (Mus Galapagoensis), and this is confined, as far as I could ascertain, to Chatham Island, the most easterly island of the group. It belongs, as I am informed by Mr. Waterhouse, to a division of the family of mice characteristic of America. At James Island, there is a rat sufficiently distinct from the common kind to have been named and described by Mr. Waterhouse; but as it belongs to the old-world division of the family, and as this island has been frequented by ships for the last hundred and fifty years, I can hardly doubt that this rat is merely a variety produced by the new and peculiar climate, food, and soil, to which it has been subjected. Although no one has a right to speculate without distinct facts, yet even with respect to the Chatham Island mouse, it should be borne in mind, that it may possibly be an American species imported here; for I have seen, in a most unfrequented part of the Pampas, a native mouse living in the roof of a newly built hovel, and therefore its transportation in a vessel is not improbable: analogous facts have been observed by Dr. Richardson in North America.

Of land-birds I obtained twenty-six kinds, all peculiar to the group and found nowhere else, with the exception of one lark-like finch from North America (Dolichonyx oryzivorus), which ranges on that continent as far north as $54°$, and generally frequents marshes. The other twenty-five birds consist, firstly, of a hawk, curiously intermediate in structure between a buzzard and the American group of carrion-feeding Polybori; and with these latter birds it agrees most closely in every habit and even tone of voice. Secondly, there are two owls, representing the short-eared and white barn-owls of Europe. Thirdly, a wren, three tyrant-flycatchers (two of them species of Pyrocephalus, one or both of which would be ranked by some ornithologists as only varieties), and a dove — all analogous to, but distinct from, American species. Fourthly, a swallow, which though differing from the Progne purpurea of both Americas, only in being rather duller colored, smaller, and slenderer, is considered by Mr. Gould as specifically distinct. Fifthly, there are three species of mocking thrush — a form highly characteristic of America. The remaining land-birds form a most singular group of finches, related to each other in the structure of their beaks, short tails, form of body and plumage: there are thirteen species, which Mr. Gould has divided into four sub-groups. All these species are peculiar to this archipelago; and so is the whole group, with the exception of one species of the sub-group Cactornis, lately brought from Bow Island, in the Low Archipelago. Of Cactornis, the two species may be often seen climbing about the flowers of the cactus-trees; but all the other species of this group of finches, mingled together in flocks, feed on the dry and sterile ground of the lower districts. The males of all, or certainly of the greater number, are jet black; and the females (with perhaps one or two exceptions) are brown. The most curious fact is the perfect gradation in the size of the beaks in the different species of Geospize, from one as large as that of a hawfinch to that of a chaffinch, and (if Mr. Gould is right in including his sub-group, Certhidea, in the main group) even to that of a warbler. The largest beak in the genus Geospiza is shown in Fig. 1, and the smallest in Fig. 3; but instead of there being only one intermediate species, with a beak of the size shown in Fig. 2, there are no less than six species with insensibly graduated beaks. The beak of the sub-group Certhidea, is shown in Fig. 4. The beak of Cactornis is somewhat like that of a starling; and that of the fourth sub-group, Camarhynchus, is slightly parrot-shaped. Seeing this gradation and diversity of structure in one small, intimately related group of birds, one might really fancy that from an original paucity of birds in this archipelago, one species had been taken and modified for different ends. In a like manner it might be fancied that a bird originally a buzzard, had been induced here to undertake the office of the carrion-feeding Polybori of the American continent.

Of waders and water-birds I was able to get only eleven kinds, and of these

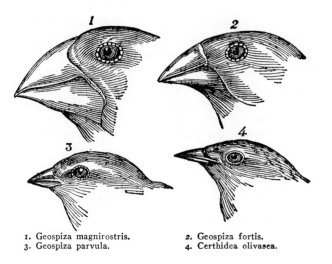

1. Geospiza magnirostris.   2. Geospiza fortis.
3. Geospiza parvula.   4. Certhidea olivasea.

only three (including a rail confined to the damp summits of the islands) are new species. Considering the wandering habits of the gulls, I was surprised to find that the species inhabiting these islands is peculiar, but allied to one from the southern parts of South America. The far greater peculiarity of the land-birds, namely, twenty-five out of twenty-six, being new species, or at least new races, compared with the waders and web-footed birds, is in accordance with the greater range which these latter orders have in all parts of the world. We shall hereafter see this law of aquatic forms, whether marine or freshwater, being less peculiar at any given point of the earth's surface than the terrestrial forms of the same classes, strikingly illustrated in the shells, and in a lesser degree in the insects of this archipelago.

Two of the waders are rather smaller than the same species brought from other places: the swallow is also smaller, though it is doubtful whether or not it is distinct from its analogue. The two owls, the two tyrant-catchers (Pyrocephalus) and the dove, are also smaller than the analogous but distinct species, to which they are most nearly related; on the other hand, the gull is rather larger. The two owls, the swallow, all three species of mocking-thrush, the dove in its separate colours though not in its whole plumage, the Totanus, and the gull, are likewise duskier coloured than their analogous species; and in the case of the mockingthrush and Totanus, than any other species of the two genera. With the exception of a wren with a fine yellow breast, and of a tyrant-flycatcher with a scarlet tuft and breast, none of the birds are brilliantly coloured, as might have been expected in an equatorial district. Hence it would appear probable, that the same causes which here make the immigrants of some peculiar species smaller, make most of the peculiar Galapageian species also smaller, as well as very generally more dusky coloured. All the plants have a wretched, weedy appearance, and I did not see one beautiful flower. The insects, again, are small-sized and dull-coloured, and, as Mr. Waterhouse informs me, there is nothing in their general appearance which would have led him to imagine that they had come from under the equator.[1] The birds, plants, and insects have a desert character, and are not more brilliantly coloured than those from southern Patagonia; we may, therefore, conclude that the usual gaudy colouring of the inter-tropical productions, is not related either to the heat or light of those zones, but to some other cause, perhaps to the conditions of existence being generally favourable to life.

We will now turn to the order of reptiles, which gives the most striking character to the zoology of these islands. The species are not numerous, but the numbers of individuals of each species are extraordinarily great. There is one small lizard belonging to a South American genus, and two species (and probably more) of the Amblyrhynchus — a genus confined to the Galapagos Islands. There is one snake which is numerous; it is identical, as I am informed by M. Bibron, with the Psammophis Temminckii from Chile.[2] Of seaturtle I believe there are more than one species; and of tortoises there are, as we shall presently show, two or three species or races. Of toads and frogs there are none: I was surprised at this, considering how well suited for them the temperate and damp upper woods appeared to be. It recalled to my mind the remark made by Bory St. Vincent,[3] namely, that none of this family are found on any of the volcanic islands in the great oceans. As far as I can ascertain from various works, this seems to hold good throughout the Pacific, and even in the large islands of the Sandwich archipelago. Mauritius offers an apparent exception, where I saw the Rana Mascariensis in abundance: this frog is said now to inhabit the Seychelles, Madagascar, and Bourbon; but on the other hand, Du Bois, in his voyage in 1669, states that there were no reptiles in Bourbon except tortoises; and the Officer du Roi asserts that before 1768 it had been attempted, without success, to introduce frogs into Mauritius — I presume for the purpose of eating: hence it may be well doubted whether this frog is an aboriginal of these islands. The absence of the frog family in the oceanic islands is the more remarkable, when

[1] The progress of research has shown that some of these birds, which were then thought to be confined to the islands, occur on the American continent. The eminent ornithologist, Mr. Sclater, informs me that this is the case with the Strix punctatissima and Pyrocephalus nanus; and probably with the Otus Galapagoensis and Zenaida Galapagoensis: so that the number of endemic birds is reduced to twenty-three, or probably to twenty-one. Mr. Sclater thinks that one or two of these endemic forms should be ranked rather as varieties than species, which always seemed to me probable.

[2] This is stated by Dr. Günther (Zoolog. Soc., Jan. 24th, 1859) to be a peculiar species, not known to inhabit any other country.

[3] Voyage aux Quatre Iles d'Afrique. With respect to the Sandwich Islands, see Tyerman and Bennett's Journal, vol. i. p. 434. For Mauritius, see Voyage par un Officier, etc., part i. p. 170. There are no frogs in the Canary Islands (Webb et Berthelot, Hist. Nat. des Iles Canaries). I saw none at St. Jago in the Cape de Verds. There are none at St. Helena.

contrasted with the case of lizards, which swarm on most of the smallest islands. May this difference not be caused, by the greater facility with which the eggs of lizards, protected by calcareous shells, might be transported through salt-water, than could the slimy spawn of frogs?

I will first describe the habits of the tortoise (Testudo nigra, formerly called Indica), which has been so frequently alluded to. These animals are found, I believe, on all the islands of the archipelago; certainly on the greater number. They frequent in preference the high damp parts, but they likewise live in the lower and arid districts. I have already shown, from the numbers which have been caught in a single day, how very numerous they must be. Some grow to an immense size: Mr. Lawson, an Englishman, and vice-governor of the colony, told us that he had seen several so large, that it required six or eight men to lift them from the ground; and that some had afforded as much as two hundred pounds of meat. The old males are the largest, the females rarely growing to so great a size: the male can readily be distinguished from the female by the greater length of its tail. The tortoises which live on those islands where there is no water, or in the lower and arid parts of the others, feed chiefly on the succulent cactus. Those which frequent the higher and damp regions, eat the leaves of various trees, a kind of berry (called guayavita) which is acid and austere, and likewise a pale green filamentous lichen (Usnera plicata), that hangs from the boughs of the trees.

The tortoise is very fond of water, drinking large quantities, and wallowing in the mud. The larger islands alone possess springs, and these are always situated towards the central parts, and at a considerable height. The tortoises, therefore, which frequent the lower districts, when thirsty, are obliged to travel from a long distance. Hence broad and well-beaten paths branch off in every direction from the wells down to the sea-coast; and the Spaniards by following them up, first discovered the watering-places. When I landed at Chatham Island, I could not imagine what animal travelled so methodically along well-chosen tracks. Near the springs it was a curious spectacle to behold many of these huge creatures, one set eagerly travelling onwards with outstretched necks, and another set returning, after having drunk their fill. When the tortoise arrives at the spring, quite regardless of any spectator, he buries his head in the water above his eyes, and greedily swallows great mouthfuls, at the rate of about ten in a minute. The inhabitants say each animal stays three or four days in the neighbourhood of the water, and then returns to the lower country; but they differed respecting the frequency of these visits. The animal probably regulates them according to the nature of the food on which it has lived. It is, however, certain, that tortoises can subsist even on these islands where there is no other water than what falls during a few rainy days in the year.

I believe it is well ascertained, that the bladder of the frog acts as a reservoir for the moisture necessary to its existence: such seems to be the case with the tortoise. For some time after a visit to the springs, their urinary bladders are distended with fluid, which is said gradually to decrease in volume, and to become less pure. The inhabitants, when walking in the lower district, and overcome with thirst, often take advantage of this circumstance, and drink the contents of the bladder if full: in one I saw killed, the fluid was quite limpid, and had only a very slightly bitter taste. The inhabitants, however, always first drink the water in the pericardium, which is described as being best.

The tortoises, when purposely moving towards any point, travel by night and day, and arrive at their journey's end much sooner than would be expected. The inhabitants, from observing marked individuals, consider that they travel a distance of about eight miles in two or three days. One large tortoise, which I watched, walked at the rate of sixty yards in ten minutes, that is 360 yards in the hour, or four miles a day, — allowing a little time for it to eat on the road. During the breeding season, when the male and female are together, the male utters a hoarse roar or bellowing, which, it is said, can be heard at the distance of more than a hundred yards. The female never uses her voice, and the male only at these times; so that when the people hear this noise, they know that the two are together. They were at this time (October) laying their eggs. The female, where the soil is sandy, deposits them together, and covers them up with sand; but where the ground is rocky she drops them indiscriminately in any hole: Mr. Bynoe found seven placed in a fissure. The egg is white and spherical; one which I measured was seven inches and three-eighths in circumference, and therefore larger than a hen's egg. The young tortoises, as soon as they are hatched, fall a prey in great numbers to the carrion-feeding buzzard. The old ones seem generally to die from accidents, as from falling down precipices: at least, several of the inhabitants told me, that they never found one dead without some evident cause.

The inhabitants believe that these animals are absolutely deaf; certainly they do not overhear a person walking close behind them. I was always amused when overtaking one of these great monsters, as it was quietly pacing along, to see how suddenly, the instant I passed, it would draw in its head and legs, and uttering a deep hiss fall to the ground with a heavy sound, as if struck dead. I frequently got on their backs, and then giving a few raps on the hinder part of their shells, they would rise up and walk away; — but I found it very difficult to keep my balance. The flesh of this animal is largely employed, both fresh and salted; and a beautifully clear oil is prepared from the fat. When a tortoise is caught, the man makes a slit in the skin near its tail, so as to see inside its body, whether the fat under the dorsal plate is thick. If it is not, the animal is liberated and it is said to recover soon from this strange operation. In order to secure the tortoise, it is not sufficient to turn them like turtle, for they are often able to get on their legs again.

There can be little doubt that this tortoise is an aboriginal inhabitant of the Galapagoes; for it is found on all, or nearly all, the islands, even on some of the smaller ones where there is no water; had it been an imported species, this would hardly have been the case in a group which has been so little frequented. Moreover, the old Bucaniers found this tortoise in greater numbers even than at present: Wood and Rogers also, in 1708, say that it is the opinion of the Spaniards, that it is found nowhere else in this quarter of the world. It is now widely distributed; but it may be questioned whether it is in any other place an aboriginal. The bones of a tortoise at Mauritius, associated with those of the extinct Dodo, have generally been considered as belonging to this tortoise; if this had been so, undoubtedly it must have been there indigenous; but M. Bibron informs me that he believes that it was distinct, as the species now living there certainly is.

The Amblyrhynchus, a remarkable genus of lizards, is confined to this archipelago; there are two species, resembling each other in general form, one being terrestrial and the other aquatic. This latter species (A. cristatus) was first characterized by Mr. Bell, who well foresaw, from its short, broad head, and strong claws of equal length, that its habits of life would turn out very peculiar, and different from those of its nearest ally, the Iguana. It is extremely common on all the islands throughout the group, and lives exclusively on the rocky sea-beaches, being never found, at least I never saw one, even ten yards in-shore.

Amblyrhynchus cristatus. *a*, Tooth of, natural size, and likewise magnified.

It is a hideous-looking creature, of a dirty black colour, stupid, and sluggish in its movements. The usual length of a full-grown one is about a yard, but there are some even four feet long; a large one weighed twenty pounds: on the island of Albemarle they seem to grow to a greater size than elsewhere. Their tails are flattened sideways, and all four feet partially webbed. They are occasionally seen some hundred yards from the shore, swimming about; and Captain Collnett, in his Voyage says, "They go to sea in herds a-fishing, and sun themselves on the rocks; and may be called alligators in miniature." It must not, however, be supposed that they live on fish. When in the water this lizard swims with perfect ease and quickness, by a serpentine movement of its body and flattened tail — the legs being motionless and closely collapsed on its sides. A seaman on board sank one, with a heavy weight attached to it, thinking thus to kill it directly; but when, an hour afterwards, he drew up the line, it was quite active.

Their limbs and strong claws are admirably adapted for crawling over the rugged and fissured masses of lava, which everywhere form the coast. In such situations, a group of six or seven of these hideous reptiles may oftentimes be seen on the black rocks, a few feet above the surf, basking in the sun with outstretched legs.

I opened the stomachs of several, and found them largely distended with minced sea-weed (Ulvae), which grows in thin foliaceous expansions of a bright green or a dull red colour. I do not recollect having observed this sea-weed in any quantity on the tidal rocks; and I have reason to believe it grows at the bottom of the sea, at some little distance from the coast. If such be the case, the object of these animals occasionally going out to sea is explained. The stomach contained nothing but the sea-weed. Mr. Baynoe, however, found a piece of crab in one; but this might have got in accidentally, in the same manner as I have seen a caterpillar, in the midst of some lichen, in the paunch of a tortoise. The intestines were large, as in other herbivorous animals. The nature of this lizard's food, as well as the structure of its tail and feet, and the fact of its having been seen voluntarily swimming out at sea, absolutely prove its aquatic habits; yet there is in this respect one strange anomaly, namely, that when frightened it will not enter the water. Hence it is easy to drive these lizards down to any little point overhanging the sea, where they will sooner allow a person to catch hold of their tails than jump into the water. They do not seem to have any notion of biting; but when much frightened they squirt a drop of fluid from each nostril. I threw one several times as far as I could, into a deep pool left by the retiring tide; but it invariably returned in a direct line to the spot where I stood. It swam near the bottom, with a very graceful and rapid movement, and occasionally aided itself over the uneven ground with its feet. As soon as it arrived near the edge, but still being under water, it tried to conceal itself in the tufts of sea-weed, or it entered some crevice. As soon as it thought the danger was past, it crawled out on the dry rocks, and shuffled away as quickly as it could. I several times caught this same lizard, by driving it down to a point, and though possessed of such perfect powers of diving and swimming, nothing would induce it to enter the water; and as often as I threw it in, it returned in the manner above described. Perhaps this singular piece of apparent stupidity may be accounted for by the circumstance, that this reptile has no enemy whatever on shore, whereas at sea it must often fall a prey to the numerous sharks. Hence, probably, urged by a fixed and hereditary instinct that the shore is its place of safety, whatever the emergency may be, it there takes refuge.

During our visit (in October), I saw extremely few small individuals of this species, and none I should think under a year old. From this circumstance it seems probable that the breeding season had not then commenced. I asked several of the inhabitants if they knew where it laid its eggs: they said that they knew nothing of its propagation, although well acquainted with the eggs of the land kind — a fact, considering how very common this lizard is, not a little extraordinary.

We will now turn to the terrestrial species (A. Demarlii), with a round tail, and toes without webs. This lizard, instead of being found like the other on all the islands, is confined to the central part of the archipelago, namely to Albemarle, James, Barrington, and Indefatigable islands. To the southward, in Charles, Hood, and Chatham islands, and to the northward, in Towers, Bindloes, and Abingdon, I neither saw nor heard of any. It would appear as if it had been created in the centre of the archipelago, and thence had been dispersed only to a certain distance. Some of these lizards inhabit the high and damp parts of the islands, but they are much more numerous in the lower and sterile districts near the coast. I cannot give a more forcible proof of their numbers, than by stating that when we were left at James Island, we could not for some time find a spot free from their burrows on which to pitch our single tent. Like their brothers the sea-kind, they are ugly animals, of a yellowish orange beneath, and of a brownish red colour above: from their low facial angle they have a singularly stupid appearance. They are, perhaps, of a rather less size than the marine species; but several of them weighed between ten and fifteen pounds. In their movements they are lazy and half torpid. When not frightened, they slowly crawl along with their tails and bellies dragging on the ground. They often stop, and doze for a minute or two, with closed eyes and hind legs spread out on the parched soil.

They inhabit burrows, which they sometimes make between fragments of lava, but more generally on level patches of the soft sandstone-like tuff. The holes do not appear to be very deep, and they enter the ground at a small angle; so that when walking over these lizard-warrens, the soil is constantly giving way, much to the annoyance of the tired walker. This animal, when making its burrow, works alternately the opposite sides of its body. One front leg for a short time scratches up the soil, and throws it towards the hind foot, which is well placed so as to heave it beyond the mouth of the hole. That side of the body being tired, the other takes up the task, and so on alternately. I watched one for a long time, till half its body was buried; I then walked up and pulled it by the tail; at this it was greatly astonished, and soon shuffled up to see what was the matter; and then stared me in the face, as much as to say, "What made you pull my tail?"

They feed by day, and do not wander far from their burrows; if frightened, they rush to them with a most awkward gait. Except when running down hill, they cannot move very fast, apparently from the lateral position of their legs. They are not at all timorous: when attentively watching any one, they curl their tails, and, raising themselves on their front legs, nod their heads vertically, with a quick movement, and try to look very fierce; but in reality they are not at all so: if one just stamps on the ground, down go their tails, and off they shuffle as quickly as they can. I have frequently observed small fly-eating lizards, when watching anything, nod their heads in precisely the same manner; but I do not at all know for what purpose. If this Amblyrhynchus is held and plagued with a stick, it will bite it very severely; but I caught many by the tail, and they never tried to bite me. If two are placed on the ground and held together, they will fight, and bite each other till blood is drawn.

The individuals, and they are the greater number, which inhabit the lower country, can scarcely taste a drop of water throughout the year; but they consume much of the succulent cactus, the branches of which are occasionally broken off by the wind. I several times threw a piece to two or three of them when together; and it was amusing enough to see them trying to seize and carry it away in their mouths, like so many hungry dogs with a bone. They eat very deliberately, but do not chew their food. The little birds are aware how harmless these creatures are: I have seen one of the thick-billed finches picking at one end of a piece of cactus (which is much relished by all the animals of the lower region), whilst a lizard was eating at the other end; and afterwards the little bird with the utmost indifference hopped on the back of the reptile.

I opened the stomachs of several, and found them full of vegetable fibres and leaves of different trees, especially of an acacia. In the upper region they live chiefly on the acid and astringent berries of the guayavita, under which trees I have seen these lizards and the huge tortoises feeding together. To obtain the acacia-leaves they crawl up the low stunted trees; and it is not uncommon to see a pair quietly browsing, whilst seated on a branch several feet above the ground. These lizards, when cooked, yield a white meat, which is liked by those whose stomachs soar above all prejudices.

Humboldt has remarked that in intertropical South America, all lizards which inhabit dry regions are esteemed delicacies for the table. The inhabitants state that those which inhabit the upper damp parts drink water, but that the others do not, like the tortoises, travel up for it from the lower sterile country. At the time of our visit, the females had within their bodies numerous, large, elongated eggs, which they lay in their burrows: the inhabitants seek them for food.

These two species of Amblyrhynchus agree, as I have already stated, in their general structure, and in many of their habits. Neither have that rapid movement, so characteristic of the genera Lacerta and Iguana. They are both herbivorous, although the kind of vegetation on which they feed is so very different. Mr. Bell has given the name to the genus from the shortness of the snout: indeed, the form of the mouth may almost be compared to that of the tortoise: one is led to suppose that this is an adaptation to their herbivorous appetites. It is very interesting thus to find a well-characterized genus, having its marine and terrestrial species, belonging to so confined a portion of the world. The aquatic species is by far the most remarkable, because it is the only existing lizard which lives on marine vegetable productions. As I at first observed, these islands are not so remarkable for the number of the species of reptiles, as for that of the individuals; when we remember the well-beaten paths made by the thousands of huge tortoises — the many turtles — the great warrens of the terrestrial Amblyrhynchus — and the groups of the marine species basking on the coast-rocks of every island — we must admit that there is no other quarter of the world where this Order replaces the herbivorous mammalia in so extraordinary a manner. The geologist on hearing this will probably refer back in his mind to the Secondary epochs, when lizards, some herbivorous, some carnivorous, and of dimensions comparable only with our existing whales, swarmed on the land and in the sea. It is, therefore, worthy of his observation, that this archipelago, instead of possessing a humid climate and rank vegetation, cannot be considered otherwise than extremely arid, and, for an equatorial region, remarkably temperate.

To finish with the zoology: the fifteen kinds of sea-fish which I procured here are all new species; they belong to twelve genera, all widely distributed, with the exception of Prionotus, of which the four previously known species live on the eastern side of America. Of land-shells I collected sixteen kinds (and two marked varieties), of which, with the exception of one Helix found at Tahiti, all are peculiar to this archipelago: a single fresh-water shell (Paludina) is common to Tahiti and Van Diemen's Land. Mr. Cuming, before our voyage, procured here ninety species of sea-shells and this does not include several species not yet specifically examined, of Trochus, Turbo, Monodonta, and Nassa. He has been kind enough to give me the following interesting results: Of the ninety shells, no less than forty-seven are unknown elsewhere — a wonderful fact, considering how widely distributed sea-shells generally are. Of the forty-three shells found in other parts of the world, twenty-five inhabit the western coast of America, and of these eight are distinguishable as varieties; the remaining eighteen (including one variety) were found by Mr. Cuming in the Low Archipelago, and some of them also at the Philippines. This fact of shells from islands in the central parts of the Pacific occurring here, deserves notice, for not one single sea-shell is known to be common to the islands of that ocean and to the west coast of America. The space of open sea running north and south off the west coast, separates two quite distinct conchological provinces; but at the Galapagos Archipelago we have a halting-place, where many new forms have been created, and whither these two great conchological provinces have each sent up several colonists. The American province has also sent here representative species; for there is a Galapageian species of Monoceros, a genus only found on the west coast of America; and there are Galapageian species of Fissurella and Cancellaria, genera common on the west coast, but not found (as I am informed by Mr. Cuming) in the central islands of the Pacific. On the other hand, there are Galapageian species of Oniscia and Stylifer, genera common to the West Indies and to the Chinese and Indian seas, but not found either on the west coast of America or in the central Pacific. I may here add, that after the comparison by Messrs. Cuming and Hinds of about 2000 shells from the eastern and western coasts of America, only one single shell was found in common, namely, the Purpura patula, which inhabits the West Indies, the coast of Panama, and the Galapagos. We have, therefore, in this quarter of the world, three great conchological sea-provinces, quite distinct, though surprisingly near each other, being separated by long north and south spaces either of land or of open sea.

I took great pains in collecting the insects, but excepting Tierra del Fuego, I never saw in this respect so poor a country. Even in the upper and damp region I procured very few, excepting some minute Diptera and Hymenoptera, mostly of common mundane forms. As before remarked, the insects, for a tropical region, are of very small size and dull colours. Of beetles I collected twenty-five species (excluding a Dermestes and Corynetes imported, wherever a ship touches); of these, two belong to the Harpalidae, two to the Hydrophilidae, nine to three families of the Heteromera, and the remaining twelve to as many different families. This circumstance of insects (and I may add plants), where few in number, belonging to many different families, is, I believe, very general. Mr. Waterhouse, who has published[4] an account of the insects of this archipelago, and to whom I am indebted for the above details, informs me that there are several new genera: and that of the genera not new, one or two are American, and the rest of mundane distribution. With the exception of a wood-feeding Apate, and of one or probably two water-beetles from the American continent, all the species appear to be new.

The botany of this group is fully as interesting as the zoology. Dr. J. Hooker will soon publish in the "Linnean Transactions" a full account of the Flora, and I am much indebted to him for the following details. Of flowering plants there are, as far as at present is known, 185 species, and 40 cryptogamic species, making altogether 225; of this number I was fortunate enough to bring home 193. Of the flowering plants, 100 are new species, and are probably confined to this archipelago. Dr. Hooker conceives that, of the plants not so confined, at least 10 species found near the cultivated ground at Charles Island, have been imported. It is, I think, surprising that more American species have not been introduced naturally, considering that the distance is only between 500 and 600 miles from the continent; and that (according to Collnet, p. 58) drift-wood, bamboos, canes, and the nuts of a palm, are often washed on the south-eastern shores. The proportion of 100 flowering plants out of 185 (or 175 excluding the imported weeds) being new, is sufficient, I conceive, to make the Galapagos Archipelago a distinct botanical province; but this Flora is not nearly so peculiar as that of St. Helena, nor, as I am informed by Dr. Hooker, of Juan Fernandez. The peculiarity of the Galapageian Flora is best shown in certain families: — thus there are 21 species of Compositae, of which 20 are peculiar to this archipelago; these belong to twelve genera, and of these genera no less than ten are confined to the archipelago! Dr. Hooker informs me that the Flora has an undoubtedly Western American character; nor can he detect in it any affinity with that of the Pacific. If, therefore, we except the eighteen marine, the one fresh-water, and one land-shell, which have apparently come here as colonists from the central islands of the Pacific, and likewise the one distinct Pacific species of the Galapageian group of finches, we see that this archipelago, though standing in the Pacific Ocean, is zoologically part of America.

[4] Ann. and Mag. of Nat. Hist., vol. xvi, p. 19.

If this character were owing merely to immigrants from America, there would be little remarkable in it; but we see that a vast majority of all the land animals, and that more than half of the flowering plants, are aboriginal productions. It was most striking to be surrounded by new birds, new reptiles, new shells, new insects, new plants, and yet by innumerable trifling details of structure, and even by the tones of voice and plumage of the birds, to have the temperate plains of Patagonia, or rather the hot dry deserts of Northern Chile, vividly brought before my eyes. Why, on these small points of land, which within a late geological period must have been covered by the ocean, which are formed by basaltic lava, and therefore differ in geological character from the American continent, and which are placed under a peculiar climate, — why were their aboriginal inhabitants, associated, I may add, in different proportions both in kind and number from those on the continent, and therefore acting on each other in a different manner — why were they created on American types of organization? It is probable that the islands of the Cape de Verd group resemble, in all their physical conditions, far more closely the Galapagos Islands, than these latter physically resemble the coast of America, yet the aboriginal inhabitants of the two groups are totally unlike; those of the Cape de Verd Islands bearing the impress of Africa, as the inhabitants of the Galapagos Archipelago are stamped with that of America.

I have not as yet noticed by far the most remarkable feature in the natural history of this archipelago; it is, that the different islands to a considerable extent are inhabited by a different set of beings. My attention was first called to this fact by the Vice-Governor, Mr. Lawson, declaring that the tortoises differed from the different islands, and that he could with certainty tell from which island any one was brought. I did not for some time pay sufficient attention to this statement, and I had already partially mingled together the collections from two of the islands. I never dreamed that islands, about 50 or 60 miles apart, and most of them in sight of each other, formed of precisely the same rocks, placed under a quite similar climate, rising to a nearly equal height, would have been differently tenanted; but we shall soon see that this is the case. It is the fate of most voyagers, no sooner to discover what is most interesting in any locality, than they are hurried from it; but I ought, perhaps, to be thankful that I obtained sufficient materials to establish this most remarkable fact in the distribution of organic beings.

The inhabitants, as I have said, state that they can distinguish the tortoises from the different islands; and that they differ not only in size, but in other characters. Captain Porter has described[5] those from Charles and from the nearest island to it, namely, Hood Island, as having their shells in front thick and turned up like a Spanish saddle, whilst the tortoises from James Island are

[5] Voyage in the U. S. ship Essex, vol. i. p. 215.

rounder, blacker, and have a better taste when cooked. M. Bibron, moreover, informs me that he has seen what he considers two distinct species of tortoise from the Galapagos, but he does not know from which islands. The specimens that I brought from three islands were young ones: and probably owing to this cause neither Mr. Gray nor myself could find in them any specific differences. I have remarked that the marine Amblyrhynchus was larger at Albemarle Island than elsewhere; and M. Bibron informs me that he has seen two distinct aquatic species of this genus; so that the different islands probably have their representative species or races of the Amblyrhynchus, as well as of the tortoise. My attention was first thoroughly aroused, by comparing together the numerous specimens, shot by myself and several other parties on board, of the mocking-thrushes, when, to my astonishment, I discovered that all those from Charles Island belonged to one species (Mimus trifasciatus); all from Albemarle Island to M. parvulus; and all from James and Chatham Islands (between which two other islands are situated, as connecting links) belonged to M. melanotis. These two latter species are closely allied, and would by some ornithologists be considered as only well-marked races or varieties; but the Mimus trifasciatus is very distinct. Unfortunately most of the specimens of the finch tribe were mingled together; but I have strong reasons to suspect that some of the species of the sub-group Geospiza are confined to separate islands. If the different islands have their representatives of Geospiza, it may help to explain the singularly large number of the species of this sub-group in this one small archipelago, and as a probable consequence of their numbers, the perfectly graduated series in the size of their beaks. Two species of the sub-group Cactornis, and two of the Camarhynchus, were procured in the archipelago; and of the numerous specimens of these two sub-groups shot by four collectors at James Island, all were found to belong to one species of each; whereas the numerous specimens shot either on Chatham or Charles Island (for the two sets were mingled together) all belonged to the two other species: hence we may feel almost sure that these islands possess their respective species of these two sub-groups. In landshells this law of distribution does not appear to hold good. In my very small collection of insects, Mr. Waterhouse remarks, that of those which were ticketed with their locality, not one was common to any two of the islands.

If we now turn to the Flora, we shall find the aboriginal plants of the different islands wonderfully different. I give all the following results on the high authority of my friend D. J. Hooker. I may premise that I indiscriminately collected everything in flower on the different islands, and fortunately kept my collections spearate. Too much confidence, however, must not be placed in the proportional results, as the small collections brought home by some other naturalists, though in some respects confirming the results, plainly show that much remains to be done in the botany of this group: the Leguminosae, moreover, has as yet been only approximately worked out: —

Name of Island	Total No. of Species	No. of Species found in other parts of the world	No. of Species confined to the Galapagos Archipelago	No. confined to the one Island	No. of Species confined to the Galapagos Archipelago, but found on more than the one Island
James Island	71	33	38	30	8
Albemarle Island	46	18	26	22	4
Chatham Island	32	16	16	12	4
Charles Island	68	39 (or 29, if the probably imported plants be subtracted)	29	21	8

Hence we have the truly wonderful fact, that in James Island, of the thirty-eight Galapageian plants, or those found in no other part of the world, thirty are exclusively confined to this one island; and in Albemarle Island, of the twenty-six aboriginal Galapageian plants, twenty-two are confined to this one island, that is, only four are at present known to grow in the other islands of the archipelago; and so on, as shown in the above table, with the plants from Chatham and Charles Islands. This fact will, perhaps, be rendered even more striking, by giving a few illustrations: — thus, Scalesia, a remarkable arborescent genus of the Compositae, is confined to the archipelago: it has six species: one from Chatham, one from Albemarle, one from Charles Island, two from James Island, and the sixth from one of the three latter islands, but it is not known from which: not one of these six species grows on any two islands. Again, Euphorbia, a mundane or widely distributed genus, has here eight species, of which seven are confined to the archipelago, and not one found on any two islands: Acalypha and Borreria, both mundane genera, have respectively six and seven species, none of which have the same species on two islands, with the exception of one Borreria, which does occur on two islands. The species of the Compositae are particularly local; and Dr. Hooker has furnished me with several other most striking illustrations of the difference of the species on the different islands. He remarks that this law of distribution holds good both with those genera confined to the archipelago, and those distributed in other quarters of the world: in like manner we have seen that the different islands have their proper species of the mundane genus of tortoise, and of the widely distributed American genus of the mocking-thrush, as well as of two of the Galapageian sub-groups of finches, and almost certainly of the Galapageian genus Amblyrhynchus.

The distribution of the tenants of this archipelago would not be nearly so wonderful, if, for instance, one island had a mocking-thrush, and a second island some other quite distinct genus; — if one island had its genus of lizard, and a second island another distinct genus, or none whatever; — or if the different islands were inhabited, not by representative species of the same genera of plants, but by totally different genera, as does to a certain extent hold good: for, to give one instance, a large berry-bearing tree at James Island has no representative species in Charles Island. But it is the circumstance, that several of the islands possess their own species of the tortoise, mocking-thrush, finches, and numerous plants, these species having the same general habits, occupying analogous situations, and obviously filling the same place in the natural economy of this archipelago, that strikes me with wonder. It may be suspected that some of these representative species, at least in the case of the tortoise and of some of the birds, may hereafter prove to be only well-marked races; but this would be of equally great interest to the philosophical naturalist. I have said that most of the islands are in sight of each other: I may specify that Charles Island is fifty miles from the nearest part of Chatham Island, and thirty-three miles from the nearest part of Albemarle Island. Chatham Island is sixty miles from the nearest part of James Island, but there are two intermediate islands between them which were not visited by me. James Island is only ten miles from the nearest part of Albemarle Island, but the two points where the collections were made are thirty-two miles apart. I must repeat, that neither the nature of the soil, nor height of the land, nor the climate, nor the general character of the associated being, and therefore their action one on another, can differ much in the different islands. If there be any sensible difference in their climates, it must be between the Windward group (namely, Charles and Chatham Islands), and that to leeward; but there seems to be no corresponding difference in the productions of these two halves of the archipelago.

The only light which I can throw on this remarkable difference in the inhabitants of the different islands, is, that very strong currents of the sea running in a westerly and W.N.W. direction must separate, as far as transportal by the sea is concerned, the southern islands from the northern ones; and between these northern islands a strong N.W. current was observed, which must effectually separate James and Albemarle Islands. As the archipelago is free to a most remarkable degree from gales of wind, neither the birds, insects, nor lighter seeds, would be blown from island to island. And lastly, the profound depth of the ocean between the islands, and their apparently recent (in a geological sense) volcanic origin, render it highly unlikely that they were ever united; and this, probably, is a far more important consideration than any other, with respect to the geographical distribution of their inhabitants. Reviewing the facts here given, one is astonished at the amount of creative force, if such an expression may be used, displayed on these small, barren, and rocky islands; and still more so, at its diverse yet analogous action on points so near each other. I have said that the

Galapagos Archipelago might be called a satellite attached to America, but it should rather be called a group of satellites, physically similar, organically distinct, yet intimately related to each other, and all related in a marked, though much lesser degree, to the great American continent.

I will conclude my description of the natural history of these islands, by giving an account of the extreme tameness of the birds.

This dispositon is common to all the terrestrial species; namely, to the mocking-thrushes, the finches, wrens, tyrant-flycatchers, the dove, and carrion-buzzard. All of them are often approached sufficiently near to be killed with a switch, and sometimes, as I myself tried, with a cap or hat. A gun is here almost superfluous; for with the muzzle I pushed a hawk off the branch of a tree. One day, whilst lying down, a mocking-thrush alighted on the edge of a pitcher, made of the shell of a tortoise, which I held in my hand, and began very quietly to sip the water; it allowed me to lift it from the ground whilst seated on the vessel: I often tried, and very nearly succeeded, in catching these birds by their legs. Formerly the birds appear to have been even tamer than at present. Cowley (in the year 1684) says that the "Turtle-doves were so tame, that they would often alight on our hats and arms, so as that we could take them alive; they not fearing man, until such time as some of our company did fire at them, whereby they were rendered more shy." Dampier also, in the same year, says that a man in a morning's walk might kill six or seven dozen of these doves. At present, although certainly very tame, they do not alight on people's arms, nor do they suffer themselves to be killed in such large numbers. It is surprising that they have not become wilder; for these islands during the last hundred and fifty years have been frequently visited by bucaniers and whalers; and the sailors, wandering through the wood in search of tortoises, always take cruel delight in knocking down the little birds.

These birds, although now still more persecuted, do not readily become wild. In Charles Island, which had then been colonized about six years, I saw a boy sitting by a well with a switch in his hand, with which he killed the doves and finches as they came to drink. He had already procured a little heap of them for his dinner; and he said that he had constantly been in the habit of waiting by this well for the same purpose. It would appear that the birds of this archipelago, not having as yet learnt that man is a more dangerous animal than the tortoise or the Amblyrhynchus, disregard him, in the same manner as in England shy birds, such as magpies, disregard the cows and horses grazing in our fields.

The Falkland Islands offer a second instance of birds with a similar disposition. The extraordinary tameness of the little Opetiorhynchus has been remarked by Pernety, Lesson, and other voyagers. It is not, however, peculiar to that bird: the Polyborus, snipe, upland and lowland goose, thrush, bunting, and even some true hawks, are all more or less tame. As the birds are so tame there, where foxes, hawks, and owls occur, we may infer that the absence of all rapacious animals at the Galapagos, is not the cause of their tameness here. The upland geese at the Falklands show, by the precaution they take in building on the islets, that they are aware of their danger from the foxes; but they are not by this rendered wild towards man. This tameness of the birds, especially of the waterfowl, is strongly contrasted with the habits of the same species in Tierra del Fuego, where for ages past they have been persecuted by the wild inhabitants. In the Falklands, the sportsman may sometimes kill more of the upland geese in one day than he can carry home; whereas in Tierra del Fuego it is nearly as difficult to kill one, as it is in England to shoot the common wild goose.

In the time of Pernety (1763), all the birds there appear to have been much tamer than at present; he states that the Opetiorhynchus would almost perch on his finger; and that with a wand he killed ten in half an hour. At that period the birds must have been about as tame as they now are at the Galapagos. They appear to have learnt caution more slowly at these latter islands than at the Falklands, where they have had proportionate means of experience; for besides frequent visits from vessels, those islands have been at intervals colonized during the entire period. Even formerly, when all the birds were so tame, it was impossible by Pernety's account to kill the black-necked swan — a bird of passage, which probably brought with it the wisdom learnt in foreign countries.

I may add that, according to Du Bois, all the birds at Bourbon in 1571-72, with the exception of the flamingoes and geese, were so extremely tame, that they could be caught by the hand, or killed in any number with a stick. Again, at Tristan d'Acunha in the Atlantic, Carmichael[6] states that the only two land-birds, a thrush and a bunting, were "so tame as to suffer themselves to be caught with a hand-net." From these several facts we may, I think, conclude, first, that the wildness of birds with regard to man, is a particular instinct directed against him, and not dependent upon any general degree of caution arising from other sources of danger; secondly, that it is not acquired by individual birds in a short time, even when much persecuted; but that in the course of successive generations it becomes hereditary. With domesticated animals we are accustomed to see new mental habits or instincts acquired or rendered hereditary; but with animals in a state of nature, it must always be most difficult to discover instances of acquired hereditary knowledge. In regard to the wildness of birds towards man, there is no way of accounting for it, except as an inherited habit: comparatively few young birds, in any one year,

---

[6] Linn. Trans., vol. xii. p. 496. The most anomalous fact on this subject which I have met with is the wildness of the small birds in the Arctic parts of North America (as described by Richardson, Fauna Bor., vol. ii. p. 332), where they are said never to be persecuted. This case is the more strange, because it is asserted that some of the same species in their winterquarters in the United States are tame. There is much, as Dr. Richardson well remarks, utterly inexplicable connected with the different degrees of shyness and care with which birds conceal their nests. How strange it is that the English wood-pigeon, generally so wild a bird, should very frequently rear its young in shrubberies close to houses!

have been injured by man in England, yet almost all, even nestlings, are afraid of him; many individuals, on the other hand, both at the Galapagos and at the Falklands, have been pursued and injured by man, yet have not learned a salutary dread of him. We may infer from these facts, what havoc the introduction of any new beast of prey must cause in a country, before the instincts of the indigenous inhabitants have become adapted to the stranger's craft or power.

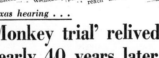

From The Changing Classroom by Arnold D. Grobman. New York: Doubleday and Company, 1969.

On the Evolution
of Insular Forms

Observations on the plant and animal life of the Galápagos islands had a tremendous influence on Darwin. It was, according to Professor Ernst Mayr of Harvard University, sometime between his visit to these islands in late 1835 and mid-1837 that Darwin became convinced of the occurrence of evolution. The personal impact of this conviction has been described by Darwin in the concluding paragraph of *Origin of Species.*

It is interesting to contemplate an entangled bank, clothed with many plants of many kinds, with birds singing in the bushes, with various insects flitting about, and with worms crawling through the damp earth, and to reflect that these elaborately constructed forms, so different from each other, and dependent on each other in so complex a manner, have all been produced by laws acting around us. These laws, taken in the largest sense, being Growth and Reproduction; Inheritance which is almost implied by reproduction; Variability from indirect and direct action of the external conditions of life, and from use and disuse; a Ratio of Increase so high as to lead to a Struggle for Life, and as a consequence to Natural Selection, entailing Divergence of Character and the Extinction of less-improved forms. Thus, from the war of nature, from famine and death, the most exhalted object which we are capable of conceiving, namely, the production of higher animals, directly follows. There is a grandeur in this view of life, with its several powers, having been originally breathed into a few forms or into one; and that, whilst this planet has gone cycling on according to the fixed law of gravity, from so simple a beginning endless forms most beautiful and most wonderful have been, and are being, evolved.

The natural history of islands is a humble topic with which to initiate a discussion of evolution. Humble? Perhaps, but also presumptuous! Have we not in this instance used as our "literary" selection nothing less than an account of evolution on the Galápagos Islands by Darwin himself? My own essay will skirt about the description and data set down by Darwin. It will instead skip from colonization of islands to the adaptation of colonizers to their insular habitat and, thence, to the consequences of repeated invasions by closely related forms.

Interest in island plants or animals is limited for the most part to the plants and animals inhabiting oceanic islands, islands rather far from continental land masses. These islands are generally of volcanic origin; they represent the tops of mountains rising from the bottom of the sea. Interest centers, moreover, on plants and animals that have a reasonably difficult time colonizing such islands. Only such inefficient migrants have an opportunity to change while isolated from their continental relatives and, hence, have an opportunity to become different.

The finches of the Galápagos Islands are one of the inhabitants that impressed Darwin at the time of his visit. Thirteen different species are to be found on the 15 islands of the Galápagos archipelago; as many as 10 of these species may be present on a large island, as few as three on a small one. These finches resemble a species of South American finch but, nevertheless, differ from the continental form. The Galápagos finches differ from one another as well. "One is astonished at the amount of creative force, if such an expression may be used, displayed on these small, barren, and rocky islands"; for one not yet committed to evolution, the puzzle was intriguing indeed.

The number of islands that are clustered in a small area is an important key to the diversity of finches on the Galápagos. Six hundred miles to the north of these islands is Cocos Island. It too has a species of finch. And this finch also differs from the South American species. It also differs from those of the Galápagos. Cocos Island is too far from the Galápagos to share even occasional migrants with them, whereas the Galápagos are sufficiently near one another that they exchange migrant finches at periodic, although rare, intervals. The population of finches on Cocos Island has undergone evolutionary change during the time it has been isolated from other comparable populations; those on the Galápagos have undergone evolutionary change both in the sense of Cocos-Island-type *change* and in the sense of the *proliferation of new species* that no longer interbreed.

Isolated oceanic islands are generally inhabited by but one species of a particular kind; archipelagos such as the Galápagos or the Hawaiian islands are centers of proliferation of related species. Probably the greatest diversity of *Drosophila* (of which the common vinegar fly is one species) has occurred within the Hawaiian islands. The different species vary a thousandfold in size, and behavioral patterns are equally diverse and bizarre.

The ways by which an animal such as a bird can make a living are not unlimited in number. It can eat either plants (herbivore) or meat (carnivore). It can catch insects or other small mammals or birds. For each way of making a living there are certain tools of the trade that are needed. Woodpeckers need a probe and a chisel. The finch on the Galápagos that hunts insects in rotten wood uses its beak as a chisel but uses a cactus thorn for a probe, whereas a real woodpecker uses its tongue as a probe. Birds that eat heavy seeds have sturdy "nutcracker" bills; as expected, the finches that eat seeds have the heaviest bills to be found among the species of finches on the Galápagos islands.

An isolated island group such as the Galápagos finches adapts as best as possible, using the genetic material at hand in evolving new ways of life and new ways of making a living. Of the finches on these islands, the one that now plays the role of a woodpecker is better at pecking wood than are any of the other types of finches and, consequently, it is successful at reproducing and maintaining itself. But the Galápagos woodpecker is a rather sorry (though interesting) woodpecker that would be crowded out of existence by nearly any continental species of woodpecker. This is a common feature of island forms. They cannot compete with similar forms that have evolved from, and in competition with, the larger array of life forms that inhabit a continent. The marsupials (pouched mammals) of Australia have evolved in diverse ways and have taken over ways of life that correspond to those of the more familiar wolves, rabbits, squirrels, moles, and other true placental mammals. With the colonization of Australia by man and the introduction of true mammals from other, larger continents, the specialized marsupials are rapidly disappearing. The wild dog, for example, has virtually displaced the Tasmanian wolf, a rather large carnivore related to the kangaroo and koala bear.

Invaders that represent successful forms of life in a much more diversified and competitive arena tend to displace their insular counterparts, for the latter have developed in isolation from a limited starting number of individuals (founders). Successful insular forms have been successful in any one instance largely because of the limited scene on which they have operated. Invaders drawn in limited numbers from approximately the same original source as the "native" insular form may at times fail to displace the "natives" because the latter have in the meantime evolved useful adaptations in response to the island environment. The result in simple cases of this sort may be the coexistence of two related species on an offshore island. Both of these species will obviously be related to a third species living on the nearby continent, but one island form will be noticeably more closely related to the continental species than will the other. This speciation pattern is characteristic of "double invasions." The first island immigrants diverge from the parental species to such a degree that, when new immigrants arrive, a reproductive barrier is built between the now co-existing island forms. These two forms diverge still further as they seek poorly exploited food sources or nesting sites within the island. Of the early and late immigrants, the second, because of the shorter period of isolation, tend to resemble more nearly the parental, continental form. Certain Tasmanian frogs occur in pairs of species which, together with related species in Australia, illustrate the outcome of double invasions of this sort.

The explosion of related species that are found in the case of the finches on the Galápagos, the honey creepers of Hawaii, or the *Drosophila* of Hawaii arises from the invasion and reinvasion of islands by immigrants from other, somewhat remote islands. Displacement following one of these invasions is possible of course; the invaders, however, are of insular origin themselves and hence not overly efficient at eking out an existence. And so, coexistence of successive waves of migrants is a common outcome. The larger the island and the more diverse its habitats, the more opportunity the various species have of finding a way of life (a "niche" in the descriptive sense of this word) that assures its continued existence. Hence, we find the 10 species of finches on one of the largest Galápagos islands but only three on some other, smaller ones. And, on Cocos Island where the invasion-reinvasion cycle is absent, only one finch species is found.

Natural Selection:
Directional and Stabilizing

To Darwin it appeared that evolution depends in large part on "a Ratio of Increase so high as to lead to a Struggle for Life, and as a consequence to Natural Selection, entailing Divergence of Character and the Extinction of less-improved forms." This view of evolution, expounded when relatively little supporting data were available from either the fossil record or from a knowledge of heredity, has remained fundamentally unchanged. The concepts of species and of species formation have been improved since Darwin's era, to be sure. Species are no longer groups of organisms that merely differ in predictable ways from other groups; rather, they are groups whose genetic material cannot be, or is unlikely to be, associated with and recombined with that of any other group. Dr. Joshua Lederberg has referred to DNA as a book in which a species writes its recipe for a successful life. A species (a "biological" species as opposed to the "taxonomic" species of Darwin) is a group of organisms that keeps its own diary.

Natural selection is no longer a vague "law." Today it is defined rather precisely so that it can be inserted into appropriate mathematical expressions. Evolution involves a change in genetic composition because in the absence of genetic change, we have only transitory alterations in the appearances of individuals. The genetic composition of a population of individuals is determined by the frequencies of contrasting alleles at various gene loci and, in the case of gene loci that are closely adjacent to one another, of their quasi-stable combinations. The rate of change in the frequency of either an allele or a specified combination of alleles can be expressed in terms that can be manipulated mathematically. Only by such manipulations can a natural law such as that of selection take on the predictive property necessary for a useful scientific law. Without predictive ability a law can only tell us what we already know.

Different patterns of selective change within populations were not clearly described – nor could they be described – in the *Origin of Species*. An alteration in the appearance or behavior of the members of a population is sufficient under the taxonomic concept of "species" to justify the naming of a new species; consequently, change and the proliferation of species went hand in hand. If, on the contrary, species are named only on the basis of their reproductive behavior vis-à-vis other groups of a similar nature, change (specifically, morphological change) has little if anything to do with speciation, with the formation of or proliferation of species. Indeed, some pairs of species are morphologically indistinguishable; such pairs are known as "sibling" species. On the other hand, a series of populations in which the individuals are clearly identifiable as to their geographical origin may go by the same species name. The English sparrow, for example, has very distinct local races in various parts of the United States; nevertheless, the species includes the total aggregate of diverse forms.

The topic for this essay is selection that produces change, that is, *directional selection*. If directional selection is not operating in respect to a given trait, the trait may be neutral or it may possess an optimal value. In the case of a neutral trait, anything goes and the trait can take on any value; in the case of a trait possessing an optimal value, aberrant individuals of both extremes tend to be eliminated from the population of reproducing individuals.

Directional selection in respect to a particular trait results in a gradual and persistent change of the population in a given direction. A population undergoing directional change is not an equilibrium population; on the contrary, it is responding to a demand of the environment. By implication, either the population has only recently entered the locality and is responding to new selective pressures, or the environment has recently undergone a change so that the population must abandon earlier characteristics and gain new ones.

The response of a population to directional selection can be demonstrated easily in the laboratory. For that matter, since the dawn of farming all kinds of agriculturally important plants and animals have undergone directional selection to better meet man's needs. Only in very recent years have exotic breeding procedures such as those used to produce hybrid corn been possible.

Evidence for the occurrence of directional selection can be seen in wild species but whether in a given instance it is still continuing can only be guessed at. I have seen on São Miguel in the Azores, for example, a blackbird scratching in the dirt and gravel at the side of a road very much the way a chicken does: this was a stouter, heavier blackbird than any I have seen elsewhere. Scratching in gravelly dirt may be a profitable occupation for an Azorean blackbird and, to whatever extent this is so, natural selection will favor heavy-bodied, stout-legged individuals that can search food among small stones more efficiently.

In most instances, directional selection does not continue to be effective for long. If a millimeter or two were added to the average height of man each generation, directional selection for increased height in man over the past 10,000 years would have increased man's height by 2 feet or more. Man's height has not increased this much during this period; those increases that have occurred have been rapid ones accompanied either by changes in nutrition, in public health measures, or in marriage patterns within and between communities.

As a rule, effective selection in a given direction ceases not because differential reproduction is eliminated but because individuals who are too tall, too heavy, or who otherwise exceed the average of the selected strain are, like those at the opposite extreme, at a reproductive disadvantage within the population. This type of selection − *stabilizing selection*, or selection for an optimal mean − is well known. The probability that a baby will survive birth, for example, is highest for babies who weigh about 7½ pounds and is lower for both heavier and lighter babies. Cultures of vinegar flies give rise to individuals with different numbers of small bristles here or there on their bodies; in those instances for which data have been obtained, flies with intermediate numbers have been found to produce the greatest numbers of offspring. Finally, in a classic study of the last century, Dr. H. C. Bumpus collected English sparrows that had been battered senseless in a violent storm. He made several measurements on the birds he collected and then found that those that recovered were of an average sort in many respects; the birds that died of exposure were more variable as a group than were the survivors.

To what sort of selection is modern man exposed? Have medical advances made selection within human populations a thing of the past? It is necessary to recall here that natural selection is no more than a reflection of differences in the reproductive success of individuals that differ genetically, provided that the genetic difference is responsible, at least in part, for reproductive success or failure. With this point disposed of, we can say that natural selection in respect to genetic differences for disease resistance has become largely ineffective (at least momentarily) because of medical advances. This statement is true only for certain populations and in respect to certain diseases; malaria is still an effective selective agent in large parts of the world. The length of time for which the statement has been even approximately true is surprisingly short − certainly less than a half-century. Modern man faces selective pressures of novel sorts. Poisons are becoming a standard feature of the environment − pesticides in the soil and water, noxious gases in the atmosphere. Noise levels are increasing. "Accidental" deaths are substantial contributors to the mortality rate. Vaccines of many sorts are being injected into ever more persons. Man will adapt to these and still other environmental and societal changes; the mechanism by which this adaptation will occur is natural selection by virtue of slight differences in the preadult mortality of different persons.

At least one scientist, Dr. René Dubos of Rockefeller University, takes a dim view of man's adaptation to a deteriorating environment.* His argument − a crystal-clear cry in an otherwise murky blend of defiance and apology − is that man can adapt to live on a garbage heap, but why should he? Today, while we still know of surroundings that are more pleasant than garbage, we should make this adaptation needless by preserving the beauty of our environment. By virtue of his Ratio of Increase, man − like all other forms of life − may be able to respond to Natural Laws such as The Struggle for Life, The Divergence of Character, and The Extinction of ill-adapted forms. Is there any need for him to do so however? This is the question that must be decided soon; otherwise, the decision will be made by default, by our collective numbers rather than our collective wisdom.

*René Dubos. Man, Medicine, and Environment. New York: New American Library (1968).

Meeting Conflicting Demands

Divergence of Character is one of the laws of nature to which Darwin referred in concluding the first edition of the *Origin of Species*.

> More living species can be supported on the same area the more they diverge in structure, habits, and constitution, of which we see proof by looking at the inhabitants of any small spot or at naturalized productions. Therefore during the modification of the descendants of any one species, and during the incessant struggle of all species to increase in numbers, the more diversified these descendants become, the better will be their chance of succeeding in the battle of life. Thus the small differences distinguishing varieties of the same species will steadily increase till they come to equal the greater differences between species of the same genus, or even of distinct genera.

Hidden within this paragraph are descriptions of at least three distinct phenomena, each of which increases the packing effect in nature, each of which leads to a greater production of living matter per unit space. I shall separate these phenomena one from the other and discuss them one by one. The first is the genetic control of development that permits the individual to take on either one of two quite different forms in response to the opportunities or demands presented by the environment. The second is the arrangement of the genetic composition of a population so that genetically diverse forms are continually produced, each of which can exploit a somewhat different environment. The third is the divergence of two forms that are already somewhat reproductively isolated so that each comes to exploit an environment or niche unlike that exploited by the other. Each of these phenomena produces a variety of types suitable for different environments, but the methods by which these are produced are fundamentally different.

The construction of two different forms by the same genetic material can be best illustrated by citing extreme examples. In a sense, however, it should not be necessary to exaggerate in this way because the plumpness, scrawniness, or robustness of persons is well known to be dependent in part upon the reaction of the individual to the amount and type of food he eats. Because campus visitors are often entertained at the Faculty Club, professors who are interviewing prospective colleagues tend to be several pounds heavier than normal; this illustrates the reaction of a given genotype to a new and richer environment.

The arrowleaf is a plant that can grow on land, partially submerged in shallow water, or completely submerged. On land the leaves are held erect on rigid stems, the root system is well developed, and the plant develops flowers. These aspects of the plant's development are needed for successful growth; nutrients must be obtained from the soil and the leaves must be oriented toward the sunlight. Underwater, the leaves are thin and ribbon-like, there are no rigid stems, and the root system is poorly developed. Again, these are useful traits for the aquatic habitat: nutrients are now obtained through the leaves, the root system is needed only to anchor the plant, and the flimsy leaves yield easily to water currents as they must if they are not to be torn off. The genotype of these contrasting forms are identical; the two developmental types represent responses of the genotype to two quite different environments.

The second example of the same sort is furnished by the life cycle of aphids (plant lice). From spring through summer, as long as plants are plentiful, these small insects reproduce asexually; wingless females produce many eggs that hatch without fertilization into new wingless females. Given a plentiful food supply, this type of reproduction is exceedingly efficient. The genotypes of all descendants of a given female are identical under this scheme of reproduction; asexual reproduction can only turn out carbon copies of the original individual.

If a plant on which aphids are living dies, the females lay eggs that produce winged females that fly away and locate new plants. Once on a fresh plant, the winged females again lay eggs that hatch into wingless females. In the fall, however, an entirely different event takes place. Eggs begin hatching into males and winged females; this ushers in a cycle of sexual reproduction, a form of reproduction that permits the creation of many, many new genotypes in contrast to the unvarying reproduction of a single genotype by the asexual process. These diverse forms of the aphid – the winged as opposed to the wingless, one versus two sexes, and the alteration between sexual and asexual modes of reproduction – are all based on the same genotype that responds in strikingly different ways to dietary, temperature, and light signals from the surrounding environment. Indeed, the utterly different larval and adult forms of butterflies, beetles, and other insects reveal the breadth of genetic control over diverse developmental patterns.

When an opportunity to pack more individuals into a series of contrasting environments arises, a species may utilize a genetic mechanism that in each generation creates two or more sorts of genetically different individuals, each of which is adapted for a particular niche in life. Perhaps the most striking examples of this solution to a difficult problem are furnished by certain African butterflies that mimic other butterfly species inhabiting the same locality. The mimics are harmless insects that could – and occasionally do – serve as food for birds and other predators; the models are ill-tasting, toxic insects that are avoided by predators. One mimic species may copy several morphologically dissimilar models. The genetic mechanism consists of a series of alleles at a particular gene locus and a hierarchy of dominance of one allele over another such that each combination of alleles resembles one or the other of the available model species. Systems such as this do not spring up full-blown overnight; a great deal of culling by natural selection is required to perfect them. Quite different color patterns are required in order to copy the different models; the wing may require a "swallow tail" in one instance and not in the other; and behavior during flight may be quite different for the mimics of one model that for those of another. These diverse aspects of the appearance and behavior of these butterflies have been physically consolidated on one chromosome and are under the control of what appears to be a single gene locus.

The third packing technique involves the evolutionary divergence of two species under the influence of natural selection. Darwin's description of divergence of character fits this case better than the two that have just been discussed. The divergence of character mentioned by Darwin is also known as "character displacement." Darwin supposed that the divergence could arise within a single species at a given locality, that is, that half of a population of birds might choose insects for food while the remainder chose seeds. A split of this sort within a single population of one species might arise as a segregating polymorphism of the sort already described for mimic butterflies. If the divergence is not of a segregating sort but represents instead a division of a breeding population into two reproductively isolated segments, it is likely that an extended period of physical isolation must precede the split. The preference for seed and insects must be developed in geographical isolation; then, upon meeting once more, the two populations can exaggerate their preferences and erect reproductive barriers so that they no longer interbreed.

Character displacement and the development of reproductive isolation are the mechanisms that lie behind the origin of the numerous finch species of the Galápagos islands. An essential feature of this proliferation of species was the existence of a number of islands far enough apart to isolate various local populations, a period of isolation during which the birds on a particular island adapted to the most promising food and the most promising way of life on that island, but close enough together to permit migrants to travel at rare intervals from one island to another. Cocos Island, which stands alone 600 miles to the north, cannot offer one segment of its finch population a period of physical isolation from the rest during which to build skills that might serve as the basis of interspecific packing. Hence, we find the single finch species on Cocos Island in contrast to the great many species that are found on the Galápagos archipelago.

● ● ●

In an earlier part of this discussion, I emphasized that the genetic mechanisms developed by the arrowleaf and aphids on the one hand and the mimic butterflies on the other operate in fundamentally different ways. The counterparts of both mechanisms are to be found in man. In the one case, individuals are genetically similar but the common genotype is such that each individual makes an adaptive response to his immediate environment. Individual plasticity offers one means for coping with the demands of a variable environment. In the other case, the individuals of the population are genetically different with each type corresponding to and adapted to one segment of a variable environment. Here the plasticity resides with the population rather than with the individual; in each generation the population produces diverse types of individuals.

Man possesses a great deal of individual plasticity; any one person can adequately carry out a wide variety of chores. Indeed, there are some who believe that all observable human variation is trivial, that all persons are basically identical, and that training alone determines a person's skills and accomplishments. Others, however, recognize that individuals are inherently different and that although each one of us exhibits considerable plasticity, those persons will perform best who are both biologically equipped and well-trained for their chosen occupations; tone-deaf persons, in a word, cannot become outstanding musicians. As an ideal goal for society, equality of opportunity makes sense only if individuals possess differing genetic endowments; it is a goal that gives the greatest opportunity for society to make the greatest use of each of its members, and for the individual members to do what they most enjoy doing. If, on the contrary, all men were biologically equal, the idea of equality of opportunity would become meaningless and, like so much of man's activities, would add nothing to either the worth of a society or the happiness of its individual members.

Instant Speciation

But, we add, these scientists have very clearly stressed that to their knowledge one does not yet know the precise meaning of the expressions "evolution," "descent," and "transition"; that one knows of no natural process by which one being could produce another of a different nature; that the process by which one species gives birth to another remains entirely impenetrable despite numerous intermediate stages; that no one has as yet succeeded experimentally in getting one species from another; and, finally, that one could never know definitely at which moment during evolution the human-like being suddenly passed the threshold of humanity.*

This essay deals with differences in the numbers of chromosomes found in related species that appear to involve multiples of some basic number (twofold, fourfold, and eightfold differences are common).† The situation I shall reveal contradicts the statement that no one has yet succeeded in getting one species from another. On the contrary! If by a species we mean a group of organisms reproductively isolated from other groups so that its genetic future depends for the most part upon changes in its own genes, then we must admit that plants have a well-known technique by which they give rise to new species almost instantaneously. Some of these have arisen within historical times and, like *Spartina townsendi*, a European marsh grass, have been extremely successful. Man has made such plant species in an effort to improve the quality and quantity of his food supply. One such man-made plant which has the leafy top of a radish and the inedible bottom of a cabbage, I must confess has been singularly useless (see diagram on page 95).

Plant species, to an extent much greater than animal species, frequently hybridize by mistake. The isolating mechanisms (procedures by which members of a species recognize and restrict their mating activities to those of their own kind) are much less absolute in the plant than in the animal kingdom. The reason probably rests in the complex nervous system of animals and the variety of recognition signals that animals can use in choosing a mate. Reproductive isolation of plant species, on the other hand, depends heavily upon such less precise things as different flowering times, pollination preferences of insects, and habitat preference.

*Pope Pius XII.

†Adapted from *Chromosomes, Giant Molecules, and Evolution* by Bruce Wallace. Illustrated by Frances Ann McKittrick. By permission of W. W. Norton & Company, Inc. Copyright © 1966 by W. W. Norton & Compamy, Inc. Additional permission by Macmillan Ltd.

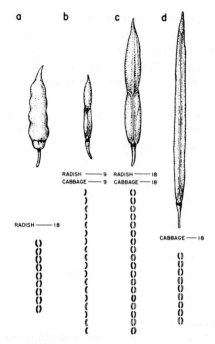

An illustration of "instant" speciation based on the radish, the cabbage, and their hybrid. Both the radish (a) and the cabbage (d) have eighteen chromosomes in cells other than germ cells. These eighteen chromosomes represent nine pairs; during germ-cell formation the halving of the chromosome number proceeds smoothly in each species. High fertility is reflected in the large seed pods.

The hybrid (b) produced by crossing the radish and the cabbage also has eighteen chromosomes; each parental germ cell contributes nine. These chromosomes are not pairs, and consequently, they do not separate regularly during germ-cell formation. The resulting imbalance in chromosome number causes a failure in the seed set of these plants; hence the seed pod is small.

If a hybrid plant is made to undergo chromosome doubling by an appropriate experimental treatment, the resulting plant (c) has thirty-six chromosomes, eighteen radish and eighteen cabbage. Germ-cell formation now proceeds regularly because each chromosome is a member of a pair, and one member of each pair goes regularly to each daughter cell during the reduction division. [After Karpechenko, based on a figure appearing in Dobzhansky, Genetics and the Origin of Species; courtesy of Columbia University Press.]

Once formed, a hybrid plant is frequently a healthy, vigorous individual (in fact, so is the mule), but it generally produces a great deal of abnormal pollen and sets very few seeds. The chromosomes the hybrid has received from each parent form a complete set so that the genetic directions for its growth and development are intact. The germ cells (eggs and sperm), on the other hand, receive only one chromosome of each pair (assuming that the chromosomes do pair in the hybrid; if they do not, the situation relative to fertility of the hybrid is worse than that described). Most of these germ cells receive an assortment of maternal and paternal chromosomes. Now, if the chromosomes contributed by the two parents (members of different species, remember) do not carry precisely the same genetic information in each chromosome of the entire set, combinations of maternal and paternal chromosomes will not be genetically complete.

We can imagine as an analogy a pair of identical twins, each of whom has packed her entire wardrobe in seven different sized suitcases. The sets of suitcases owned by the twins are identical. The two travel together on a European holiday, sharing outfits and combining items of apparel with complete abandon. This period of joint travel corresponds to the normal development of a hybrid plant. Finally, the twins decide to go separate ways. Each packs her belongings into her seven suitcases again. At the moment of parting, however, the hotel porter sends with each one a set of seven bags that is part hers part her twin's. Unless they packed exactly the same types of clothing into each bag of the set, it is possible that when the two set off, neither twin will have a fully coordinated outfit in her possession. A precisely similar difficulty faces the germ cells of a hybrid plant.

Once in a great while, a cell will accidentally fail to divide after its chromosomes have already duplicated. The result is a cell with twice as many chromosomes as a normal cell. Should this accident affect a cell early in the growth of a plant, an entire branch may be composed of such polyploid cells. In a normal individual, this doubling leads to irregularities in the formation of germ cells. The orderly reduction of chromosomes number in preparation for germ-cell formation depends upon the presence of *pairs* of chromosomes. Members of each pair seek each other out, pair, and then pass separately into the two daughter cells. The presence of four chromosomes of each type in the polyploid cell leads to difficulties in the processes of pairing and orderly separation; groups of three or four come together, and then separation goes awry.

In contrast, hybrid plants that have undergone a similar accidental chromosome doubling are occasionally spared the difficulties arising from the presence of four identical chromosomes. The troubles besetting the fertility of some hybrids result, as we suggested above, from the almost total lack of pairing between chromosomes. Polyploidy, the doubling of all chromosomes, removes that source of trouble and – presto! – the hybrid is transformed into a true-breeding individual possessing one pair of every chromosome but with twice the number (or, more correctly, the sum of the two original numbers) of chromosome pairs of the two "parental" species. The new species is true-breeding; furthermore, it gives rise to infertile offspring when crossed with either parental species. (The mechanics of the procedure we have described here are not nearly as difficult as they may seem; the diagrams presented at the end of the essay may help to make clearer.)

I do not intend to catalogue the number of plant species of polyploid origin. I shall merely assert that a great many plant species, cultivated and wild, are polyploids. Even polyploids hybridize, undergo chromosome doubling, and form higher "ploids." Botanists have, by shrewd deductions, reconstructed the steps leading to the formation of certain naturally occurring polyploid species. They have identified the parental species and produced artificial hybrids in the laboratory. Finally, by the use of certain chemicals, they have doubled the chromosome numbers of the artificial hybrids. The experimentally produced polyploid hybrids look like their natural counterparts; furthermore, many experimental and natural polyploids cross freely with one another and produce fertile offspring. There are no grounds for doubting that new plant species arise in nature precisely in the manner described here and precisely as the experimental botanist proceeds in his series of laboratory crosses. The contention that no one has as yet succeeded experimentally in getting one species from another, consequently, is wrong.

Animals cannot avail themselves easily of polyploidy as a means for forming new species; chromosome numbers in related animal species do not, as a rule, differ in simple multiples. Higher animals do not generally reproduce by self-fertilization or by asexual, clonal growth (as the growth of potatoes from tubers or of African violets from cuttings). In order for an animal polyploid species to arise successfully, a complicated combination of rare events — hybridization and doubling of chromosome number — would have to occur simultaneously in each of two individuals, one male and one female, living in the same neighborhood. Furthermore, these two individuals would have to choose each other as mates in preference to normal individuals of the two parental species living in the same region. Their offspring, too, would have to prefer one another and mate, brother with sister, for a number of generations. For reasons that we need not describe here, the germ-cell formation in males and the sex-determining mechanisms would tend to malfunction. In brief, an animal species could adopt polyploidy only as a consequence of the coincidence of four, five, or six extremely rare events. When events are extremely rare, they coincide at infinitesimally small frequencies; coincident occurrences of several are rare even over time spans measured in millions of years.

On Modern Genetics
and Evolution

The original edition of the *Origin of Species* has been reprinted recently with a foreword by Professor Ernst Mayr of Harvard University. Professor Mayr comments on the remarkable durability of Darwin's ideas about evolution despite the scant supporting evidence that was available to him from various branches of biology. As a rule, the taxonomists of Darwin's time looked upon species as the products of special acts of creation; the avowed role of taxonomists was to reveal the plan of the Creator, not to construct phylogenies depicting evolutionary descent. The paleontological record was scanty indeed; a number of fossil "links" such as *Archaeopteryx* (a primitive reptile-like bird) and *Ichthyostega* (a fish-like amphibian) were unknown. Darwin, as naturalist on the *Beagle,* was himself making fundamental contributions to biogeography, the study of the distribution of life forms on earth. His contemporary, Alfred R. Wallace, whose essay on the origin of species by natural selection was published jointly with Darwin's work, was also a pioneer in biogeography. Embryology was not an advanced science. Genetics as a science was actually unknown, and the prevailing views on inheritance were quite misleading. Despite these difficulties, the edifice erected by Darwin remains, and the contributions made in the meantime by the peripheral biological sciences have supported rather than undermined Darwin's thesis.

I propose in this essay to sketch in rough outline the contributions of genetics and molecular biology to our understanding of evolution. These contributions consist of studies on chromosome structure in related species, the amino acid sequences in proteins of related species, the use of DNA as genetic material by virtually all forms of life, and the universality of the genetic code.

The fertilized egg contains one set of chromosomes that was present in it before fertilization (the "maternal" set) and a corresponding ("paternal") set brought in by the sperm. Thus, with exceptions involving sex-determination that need not concern us here, the developing individual carries within the nucleus of each cell of his body two matching sets of chromosomes, one pair of each of a number of different (*nonhomologous*, as contrasted with the homology of the two member of each pair) chromosomes.

The chromosomes of most organisms and of most tissues within the body of an individual are rather modest physical bodies that, even with the help of an excellent microscope, reveal very little structural detail. A tremendously important exception to this rule is to be found in chromosomes of the salivary gland cells of larval flies and mosquitos (including the vinegar fly and its relatives that were chosen quite by accident for intensive study by early geneticists); within these cells the individual chromosomes are not tightly coiled structures but are extended to a total length of about one millimeter. In addition, as the larvae develop, the chromosomes reproduce copies of themselves many many times so that each "chromosome" is really a bundle of 1,000 or more parallel strands. These giant chromosomes reveal a wealth of structural detail in the form of hundreds or thousands of cross bands that make the most intricate combinations of light and dark, solid and dotted, double and single bands along the length of a chromosome, which appears under a high-power microscope to be 1 or 2 feet long (see the figure on p. 103).

The many fascinating studies that can be based upon the wealth of visible detail in giant salivary gland chromosomes are not relevant to this essay; the reader must go elsewhere to learn of them. Here I merely emphasize that the two members (one maternal, one paternal) of each pair of chromosomes not only correspond band by band but also fuse so that corresponding bands of the two members of the pair form a single band that extends unbroken across the fused, double structure as shown in the drawing. The band-by-band correspondence of maternal and paternal chromosomes attests to the constancy of these intricate chromosomal patterns. The patterns would be identical, for example, in a larval vinegar fly whose father was captured in California and whose mother came from South Africa; two flies caught at such tremendous distances from one another have probably not shared a common ancestor for thousands of years.

The band-by-band pairing is essential in revealing the gross changes in chromosomal structure that occur at rare intervals in the history of a species. The changes that are important to us are *inversions,* rearrangements of genetic material that result from the breakage of a chromosome into three parts and the reinsertion of the central segment following its rotation by 180 degrees:

Original sequence	A B C D E F G
Breaks (*)	A*B C D*E F G
Rotation	A*D C B*E F G
Healing	A D C B E F G

The occurrence of inversions can be detected by a careful study of the order of bands in giant chromosomes; if regions *B C D* have a characteristic, asymmetrical sequence of bands, the inversion of this segment *(D C B)* can be detected by the reversed order of this sequence. On the other hand, if the paternal chromosome in a larva has the gene sequence *A B C D E F G* and the maternal one *A D C B E F G*, the band-by-band pairing of these two chromosomes throws the fused chromosomes into a large loop in which one chromosome runs in one direction and the other in the reverse direction. These large loops provide the best insight as to the precise position of the breakage points that are responsible for the observed inversions (see accompanying figure).

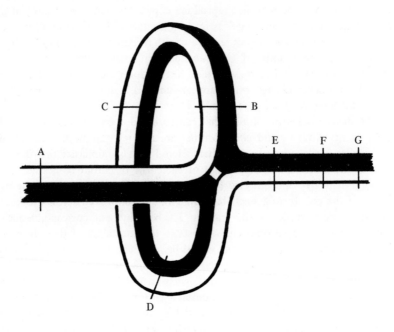

Drosophila species which are so closely related that they can hybridize and produce viable larvae have been studied in respect to chromosomal inversions that are revealed by the pairing of giant chromosomes in the hybrid larvae. The chromosomes of many different species do fuse — either completely or in substantial segments — in hybrid larvae. This suggests at once that the chromosomal material of these different species is related because only segments that correspond to one another (that are homologous) pair intimately within a single species. Second, the inversion loops detected in the hybrid larvae can be analyzed in terms of the responsible breakage points; in each case, these points fit closely with the overall pattern of chromosomal rearrangements worked out within the two species separately. Very often, related species will share one gene arrangement of a chromosome in common. In several pairs of species the second gene arrangment of a sequence of three (for example, arrangement *beta* of the sequence consisting of the three arrangments, *gamma, beta, delta* shown in the figure at the end of this essay) has been found in one species while the first and third arrangments, which demand the existence of the second, have been found in another species. Here then is elegant proof that the genetic material that is found in diverse species today must have been found in one species (indeed, one population of one species) in the past. This is the meaning of the phrase, "the origin of species."

The amino acid sequence of each protein is specified by plans carried in the responsible DNA molecule; the order of base pairs in DNA governs the order of amino acids in the protein molecule. Identical amino acid sequences in proteins of different species of plants or animals is compelling evidence that these species are related by descent. Occasionally the structural identity of two proteins may be imposed by a particular functional need; but, as a rule, the demands made by a protein's function are not so rigorous that they cannot be met in a variety of ways.

Protein analyses do indeed reveal that widely separated forms have proteins of related structure. The more remote the relationship between the species (predicted by morphological differences and other criteria), the more different the amino acid sequences in related proteins. This type of study has revealed, in addition to similarities among the proteins of different species, similarities among the proteins made by genes at different sites within the chromosomes of a single species. This finding was anticipated by early geneticists. If genes arise only from preexisting genes, then the number of genes in the entire complement of a species can increase only by the inclusion of a given gene two or more times – that is, by gene duplication. One of the two copies of the gene can then diverge (or both of them can, for that matter) in meeting new demands. Giant chromosomes reveal what appear on the basis of banding patterns to be repeated segments; analyses of amino acid sequences have revealed proteins that are now made by different genes but genes that are clearly related by descent through duplication.

That DNA is used as genetic material by virtually all forms of life (some viruses use RNA, a related chemical) may or may not be pertinent to a discussion of evolution. We know in general terms the conditions that DNA must meet to be effective as *the* genetic material: it must replicate accurately; its altered forms (mutations) must also be replicated in order to allow for inherited changes; and its structure must be transcribable so that it serves as a set of instructions for metabolic processes that are essential to life. A great deal is known about the physical chemistry of DNA and why it is suitable for the role it plays in heredity. Should no chemical other than DNA meet the demands of living systems, then the use of this substance by diverse forms tells us nothing about their evolutionary descent. With no alternative, all living forms would, under this view, of necessity possess DNA — with or without evolution.

The existence of a universal genetic code is, however, elegant evidence for the single origin of life on earth or, to be more precise, for the single origin of *currently existing* forms of life. The use of DNA might conceivably have been imposed on all newly developing living systems simply because no other chemical can serve as genetic material. But, each independent system could, at least in theory, have worked out its own genetic code. This is not what we find however. By using the protein-synthesizing machinery of one type of cell and the instructions (messenger-RNA) of another, experimentalists have shown that the resulting proteins have essentially correct amino acid sequences. There is still further evidence. Knowing that viral infections involve the introduction of alien genetic material into a cell and the preemption by viral genes of the infected cell's vital machinery, it follows that the mere existence of viruses capable of reproducing in both plants and insects, or in insects and mammals, attests to the universality of the genetic code. If the code were not universal, the same microscopic filament of DNA (or RNA) could not make the same protein by means of the machinery of an infected plant cell and, at a later time in its life cycle, by means of the machinery of an infected insect cell.

This essay on the relation of modern genetics to the theory of evolution can be concluded by the following summary.*

Rejecting as we must the untestable "explanation" of the origin of species through special creation and proposing instead (1) that freely breeding populations change (evolve) with time and (2) that reproductively isolated populations of individuals (species) arise from similar but pre-existing populations, we have proceeded to itemize that evidence from the science of genetics we considered pertinent. A century ago, in undertaking a similar task, Darwin had no alternative but to examine the geologic record, to describe paleontological finds, to study geographic distributions of related forms, and to muster the evidence provided by anatomy and embryology. The case Darwin built still stands. Here we have supplemented his observations with those drawn from genetics, cytogenetics (genetics supplemented by microscopic studies), and molecular biology. Darwin, as it were, rounded up witnesses, checked alibis, got descriptions, and established the order in which events occurred; we have checked fingerprints, blood samples, and powder burns. There is no need to discard Darwin's theory of evolution. On the contrary, as one geneticist [has written]: "The occurrence of evolution in the history of the earth is as well established as can be any event or process not witnessed by human observers, not witnessed for the simple reason that such observers did not yet exist or did not know how to record their testimony".

A drawing of the right end of chromosome 2 of the common vinegar fly, Drosophila melanogaster. *Careful examination of this figure reveals that what appears to be a single structure is really double. The two chromosomes derived from the two parents are coiled loosely around each other with corresponding bands matched perfectly through-out. There are 100 or more numbered segments in the "map" of the entire chromosome set of this species. A drawing of the entire set on the scale of the segment reproduced here would be about 2 feet in overall length. There are roughly 150 bands in the segment shown here and so, at the same degree of detail, the entire set would include about 3500 to 4000 bands.*

*Adapted from Chromosomes, Giant Molecules, and Evolution by Bruce Wallace. Illustrated by Frances Ann McKittrick. By permission of W.W. Norton & Company, Inc. Copyright c 1966 by W.W. Norton & Company, Inc.

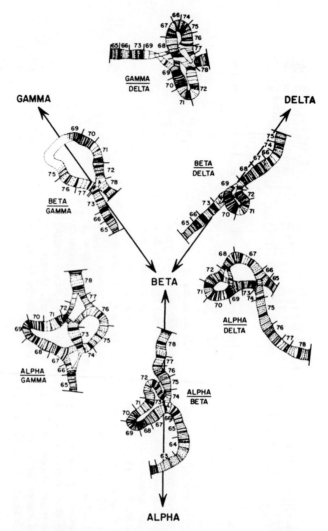

Pairing patterns of four gene arrangements found in a Mexican fruit fly, Drosophila azteca. *This figure shows the names of the different gene arrangements, their derivation one from the other, and, in addition, the pairing configurations of all combinations of the four gene arrangements. It also shows that one double-headed arrow (*beta *to* delta, *for example) stands for a single loop in the pairing of the two gene arrangements involved, while two double-headed arrows (connecting* alpha *to* delta, *for example) stand for figure-eight configurations. [After Dobzhansky and Sokolov; courtesy of the* Journal of Heredity.*]*

*Adapted from *Chromosomes, Giant Molecules, and Evolution* by Bruce Wallace. Illustrated by Frances Ann McKittrick. By permission of W.W. Norton & Company, Inc. Copyright c 1966 by W.W. Norton & Company, Inc.

The Evolution
of Ideas and Beliefs

During the past 20 years or more biology has undergone a revolution. Today is the age of molecular biology. Biological phenomena are being explained in terms undreamed of prior to World War II. At the forefront of the revolution has been a handful of physical biologists with vision and tremendous intellectual ability; behind them are hordes of specialists and bright youngsters trained in the use of new experimental techniques, equipped with new analytical devices, and confronted by new sets of fascinating problems.

Progress in science as elsewhere is not always appreciated at the moment; scepticism, dissent, and resentment can linger for decades. Problems once thought important are brushed aside; lifetimes of intellectual effort, revered at one moment, are threatened with obscurity the next. Much of the opposition and resentment against Darwin, it seems, had its origins in pique over intellectual neglect; this is the view expressed by Professor Ernst Mayr in his introduction to the Harvard reprint of *Origin of Species* :*

> For Darwin, evolution was more than change of appearance due to the unfolding of preformed inherent tendencies. His concept of evolution required a real genetic change from generation to generation, a complete break with the so-called evolutionary concepts of Lamarck and virtually all forerunners. Darwin started from a new basis by completely eliminating the last remnants of Platonism, by refusing to admit the *eidos* in any guise whatsoever. . . .
>
> No one resented Darwin's independence of thought more than the philosophers. How could anyone dare to change our concept of the universe and man's position in it without arguing for or against Plato, for or against Descartes, for or against Kant? Darwin had violated all the rules of the game by placing his argument entirely outside the traditional framework of classical philosophical concepts and terminologies. Perhaps this is the greatest difference between him and all of his predecessors, be they antievolutionists such as Linnaeus, Cuvier, and Louis Agassiz, or evolutionists such as Lamarck. No other work advertised to the world the emancipation of science from philosophy as blatantly as did Darwin's *Origin*. For this he has not been forgiven to this day by some traditional schools of philosophy. To them, Darwin is still incomprehensible, "unphilosophical," and a bête noire.

"Darwin had violated all the rules of the game by placing his argument entirely *outside the traditional framework* of classical philosophical concepts

*Reprinted from the Introduction by Ernst Mayr to Charles Darwin's *On the Origin of Species: A facsimile of the first edition* published by Harvard University Press, 1964.

and terminologies." Here, in a phrase, is the heart of a revolution. And here, too, is the basis of the revolution in modern biology: the entire machinery of living cells has been reduced to molecular gadgetry while the last possible hiding places for vitalistic forces and other mysterious life processes are being dragged into the light. In the meantime, however, problems that do not lend themselves to molecular manipulations — and persons who study such problems — are in the eyes of many molecular biologists beyond the pale. Beyond the pale? The precise interpretation of "outside the traditional framework" and "beyond the pale" hinges on the success of the revolution and on one's role vis-à-vis the uprising.

The opposition to evolution continues around the nation. One by one, to be sure, state laws against the teaching of evolution in public schools have been toppled or, in the case of those that have been proposed anew, voted down. Nevertheless, antievolutionary forces are still surprisingly powerful.

There are influential figures in the educational hierarchies of several states who urge that evolution be taught as "just a theory, not a fact." These same persons in the school-prayer controversy often claim that when religion leaves the schools, morality also leaves because there can be no morality without religion.

Within the scientific community there is no objection to the teaching of evolution as a theory. "Theory" is a respected word in science; virtually all professional views that are held by scientists are held in the form of theories. When a nonscientist proposes, however, that evolution be taught in schools as a "theory," he invariably means "just a theory" — a phrase that is to "theory" as "damn yankee" is to "yankee." "Just a theory" in pragmatic America means "wrong"; the phrase has no intellectual relationship whatsoever to the "theory" of the scientific world. Theories, to the ordinary citizen, are foolish arguments of the sort that once led scientists to conclude that bumblebees cannot fly, or they are the bizarre beliefs held by economists in the nation's capital regarding circumstances in rural America. In short, theories do not work.

Do not underestimate the claim, however, that there can be no morality without religion! Beware, that is, of believing that a system of ethics can be constructed of and erected on experimental observations; many scientists have failed in attempting to erect such a system. Darwin himself set the stage in the concluding remarks of his *Origin*. There is, as he says, grandeur in the view that the most beautiful and most wonderful forms of life have evolved from a single form. Wondrous and beautiful as the evolved forms may be, grand as the view of their origin may be, the process is still a mechanistic one. Nor has molecular biology made it any less so; molecular biology is "utterly mechanistic and the greatest triumph of the *bête machine* that Descartes could have imagined." We can claim with ample justification that every from of life on earth — plant or animal; burrowing, walking, or flying; terrestrial, arboreal, or aquatic — represents a device generated by DNA in assuring its own preservation. The body cells, the senses, and the emotions — those things about us of which both medical doctors and poets love to write — are all designed by DNA while arranging that bits and pieces of itself will be transported by eggs and sperm through time, generation by generation.

Moral laws cannot be derived from the mechanisms that preserve DNA any more than they can be sought in the procedures that enable a housewife to preserve jellies. Nevertheless, there are accepted standards of behavior in human societies. Furthermore, in Western societies these standards are as respected by agnostics as by supposedly religious persons because they have rational (not scientific) bases. "As our laws stand," wrote Thomas Huxley a century ago, "if the lowest thief steals my coat, my evidence (my opinions being known) would not be received against him." This is no longer quite the case.

In my opinion, an intelligent man explaining according to some rational scheme why he behaves "decently" does society a greater service than would a horde of flamboyant orators berating the theory of evolution for displacing religion in our schools. A logical code of ethics, one free of hypocrisy and mysticism, needs to be developed for the education of our children. It must emphasize both the equality and the immense worth of individuals. It should teach that the infliction of physical pain on, or the causation of mental anguish in, others is wrong. It should acknowledge that the individual exists from birth until death. Furthermore, it should emphasize that during his existence, a person's mind with its accumulated experiences *is the person*.

These points should be thought through soon, and they should be thought through in a systematic, intellectual way rather than on an ad hoc basis — hospital by hospital, town council by town council, school board by school board. Much more of science than Darwin's theory of evolution is battering at old beliefs; molecular studies on growth and development are threatening to lay bare the very origins of new "life." Human bodies in which only cellular processes still function are kept "alive" by heroic medical efforts. In anticipation of a future man-made "resurrection," a number of persons have already had their remains preserved by freezing. Time in which to rethink centuries of moral traditions is growing short; we can ill afford to waste any of it on reruns of Tennessee's Monkey Trial.

Technology
and the Future of Man

At a recent scientific conference, Dr. Glenn T. Seaborg of the U.S. Atomic Energy Commission took this view of the environmental crisis facing man: "What we are seeing today, in the anguish over environmental feedback and the piling of crisis on crisis, is not a forecast of doom. It is the birth pangs of a new world – the period of struggle in which we are making the physical transition from man to mankind."* These words have an impressive sound but what do they mean? What is meant by "transition from man to mankind?" I interpret such words to mean that Dr. Seaborg has a naturalistic ethic; the words "birth pangs," "struggle," and "transition" imply evolutionary change. Dr. Seaborg, if I read him correctly, believes that change is progress and, therefore, good. Man, that is, must adapt to technology in order to survive and, should he be successful, he will then qualify for the label "mankind."

Man is a moral animal. Every human society has its ethical standards; certain actions are good, others are bad. George Gaylord Simpson has pointed out that persons lacking a sense of good and bad are "felt to be as surely crippled as if the deformity were physical."† The ethics of a society are passed down from generation to generation. The ability of each individual to grasp and adhere to a set of ethical standards is presumably biological in the same sense that the physical aspects of the nervous system are under genetic control. The precise nature of ethical standards, however, is cultural; these standards form a large part of the cultural heredity of a human population.

What are the origins of a system of ethics? How do persons arrive at those beliefs that govern what is to be judged good and what bad? Such systems can be built up intuitively. Should the intuitive system devised by one person succeed in becoming a cult, the original person becomes the inspirer and his intuition may eventually be looked upon as one operating through divine revelation. Different intuitive ethics need not agree with one another; acts that are regarded as good under one system may be considered bad under another. In a search for a common ethic for man, intuition and revelation do not appear to serve as the best possible starting points.

*New York Times, May 3, 1970.
†G. G. Simpson, *The Meaning of Evolution* (New Haven: Yale University Press, 1949).

"The search for a naturalistic ethical system has had as its basis confidence in observation and experiment as leading to discovery of objective truth, or increasing approximation to it, and the conviction that what is ethically right is related in some way to what is materially true."* Even though it may be granted that an ethical system should not contradict natural truths, systems that seemingly fit this requirement need not be acceptable systems.

Evolution has served as a popular model for those (and it would appear that Dr. Seaborg is among them) who favor naturalistic ethics. To those of the last century who saw evolution as the product of raw selection — nature red in tooth and claw — a gladiatorial ethic seemed appropriate: might makes right; every man for himself. T.H. Huxley, Darwin's staunch supporter, found this notion so repugnant that he developed his own intuitive antievolutionary ethics for man. Tooth and claw was all right for nature, but man's ethical system was to be of a different sort. If, indeed, the phrase "transition of man to mankind" has a definable meaning, Huxley's intuitive ethical system would appear to oppose it. In attempting to circumvent the consequences of natural selection, Huxley found himself promoting stability for man.

In large measure, tooth-and-claw ethics have fallen into disfavor. Differential reproduction, a necessity for evolutionary change, is not always — in fact, is seldom — dependent upon the killing of one individual by another of either the same or different species. More often selection depends upon the relation between an individual and the environment into which he has been thrust by birth. This relation is most often expressed in subtle differences in numbers of offspring, but even predation is most often restricted to the culling of those individuals who have already displayed an inability to cope with their surroundings.

Evolutionary "facts" other than tooth and claw have been utilized in constructing naturalistic ethics. "Aggregate" ethics is one of the results. The passage from primitive to modern life represents, in part, the passage from simple to complex organisms: single-celled individuals to multicellular ones. Beyond the multicellular individuals is the "epi-organism," the society to which the organism belongs. Under aggregate ethics, "good" lies in the subjugation of the individual to society much as the individual tissue cell is subjugated to the needs of the body. Nazi Germany serves as the notorious example of a society in which individuals were subjugated to the larger group; few persons would care to admit that this period of German history is an enviable one or noted for its goodness.

Most persons are not practicing moralists; most of us, that is, are content to go through life acting toward others in an intuitively decent (or is it conditioned?) manner without probing by asking "Why?" Essays such as Garret Hardin's the *"Tragedy of the Commons"* are needed to jolt us into the sad

*Simpson, *The Meaning of Evolution*, p. 297.

realization that decent — even rational — behavior cannot solve today's problems. We harm others even by our decency; solutions for problems such as overpopulation, pollution, and the depletion of natural resources require a new morality.

Technologists turn (perhaps unwittingly) to naturalistic ethics in promoting their favorite projects. This fact was strikingly brought home by a panel discussion on TV following man's first moon landing. An outspoken social critic was alone in questioning the wisdom of the nation's priorities in general and of the moon shot in particular. Other members of the panel were seemingly cowed by the eminently confident representative of the national space agency who triumphantly announced that *mankind had now achieved immortality.* Our earth, he went on to explain, has a finite lifespan because its sun will eventually die; but with his landing on the moon, man has escaped the threat of extinction, for he will now populate worlds that revolve around other suns. The universe, in effect, is ours.

At the time of his appearance on the panel discussion and in a subsequent article,* this spokesman, Robert Jastrow, also justified (in part, at least) the construction of a space station by citing its use in locating schools of fish for marine fishermen. This argument more than any other reveals a woeful lack of biological knowledge on the part of space technologists. The breeding sites of marine life, the coastal marshes, have been systematically erased by man's landfills and industrial pollutants until they have all but disappeared. Modern fishing boats equipped with the latest electronic gear are already dangerously overharvesting the oceans. American fishermen are continually running afoul of the Peruvian authorities because we insist on fishing in what Peru regards as its national waters. Despite these facts, the space agency predicts an increase of 2 billion dollars in the world's annual fishing catch through the use of the space station. Should they undertake this service, I predict we shall be entering upon the terminal years of the marine fishing industry. Marine fish, like the passenger pigeon of yesterday and the whales of today, will be hounded to extinction.

Mr. Jastrow's grasp of evolutionary problems is at least as poor as his understanding of ecological ones. "The last major [evolutionary] development ... occurred 300 million years ago, when, in a time of seasonal drought, the fishes left the water and invaded the land to become the first air-breathing animal with backbones. ... Only those species equipped with stumpy fins for walking, and with lungs as well as gills for breathing, could do so. ... From those few, favored by a suitable body apparatus, and perhaps by an unusual measure of curiosity and fortitude, are descended all the air-breathing back-boned animals that now inhabit the earth." On an unseasonably hot Thursday afternoon, Mr. Jastrow seems to imply, brave and curious stumpy-finned fish

*Robert Jastrow, "Space Odyssey of Tomorrow — a Trip to Mars," *New York Times Magazine,* May 10, 1970.

decided to leave their watery home for a life on land. And that was that. The naiveté of this view of evolution is incredible. The millions of years, the gradual previous selection for those stumpy fins and rudimentary lungs, the long transitional periods of alternating rain and drought, and the uncountable multitudes of dead fish rotting in primeval swamps and mud flats have been condensed into a single moment of decision, a moment when the time for fish to leave the water had arrived. Blast-off time for the stumpy-finned ones!

Besides the abbreviated and romantic account of the transition from fish to amphibia, Jastrow presents an equally abbreviated and romantic account of the future evolution of man. In brief, those men (and women) with an unusual endowment of curiosity and fortitude set up housekeeping on Mars, a planet with "a form of life so dessicated that, when immobile, it seems to be composed of inanimate matter." That life exists on Mars is mere speculation, of course; nevertheless, this future human colony thrives (on what?), begets children, builds new rockets (with what?), and sets off beyond Mars and beyond the solar system.

Omitted from this entire superficial discussion are thorough examinations of at least two matters. (1) Why should it be necessary for man to leave earth? (2) What is the cost of natural selection? The first of these matters is mentioned only in terms of threat to life on earth: the threat of extinction in nuclear warfare, the threat of overpopulation, the threat of excess leisure time, and, by implication, the threat of an ecological catastrophe. The threat of over-population arises from an innate capacity of living things to produce more young than can normally survive. Specifically, in the case of man, the threat comes from the belief of a technological society that it can sustain on a finite globe an infinite number of persons at an unbelievably high standard of living. The other threats — nuclear warfare, excess leisure time, and ecological catastrophe — arise directly from technology itself. Jastrow transcends the technologists against whom McDermott rails.* Technology has gotten the world into this mess and it will get us out of it — not by cleaning up the mess but by transporting a small colony of excessively curious persons to Mars. And the "escape" will be effected — presto! — just as eons ago during a seasonal drought when the more curious, lobe-finned, and lunged fish came up on land to stay. Would it not be more reasonable to turn man's knowledge against overpopulation and to use his technological skills in establishing and maintaining an inhabitable earth?

The cost of natural selection does not enter into Jastrow's discussion. It did enter into T.H. Huxley's, of course, and as a result he turned away from a naturalistic ethic to an intuitive one. To what sorts of selection experiments has man been subjected lately? Several million persons were exposed to carbon monoxide and hydrogen cyanide in Germany during the early 1940's; none of those persons survived. Millions of persons have died or are dying of starvation;

*See "On the Growing Role of Technology," in Volume I.

no striking progress toward an ability to live without food has been reported. Malaria is still countered in many areas by a genetic mechanism that results in the sacrifice of nearly one-half of all children born. Selection is a slow process; progress is measured in observed gains per thousands of generations. Changes effected during a single generation are usually imperceptible. Rapid evolutionary changes such as the origin of DDT-resistance in insects or resistance to myxomatosis in Australian rabbits are accompanied not be a mere decimation of populations but by the elimination of 90 per cent, 99 per cent, or even more of all individuals. When the overpowering need arises, rapid evolution is the successful response of a minority of threatened populations; rapid extermination is a more probable outcome of such devastating selection. In either event, however, one must be inordinately naive to equate evolutionary change with desirability, to counterbalance the large-scale extermination of human beings with a claim of immortality, or to stake the questionable goodness of an aggregate ethic on the slight chance of man's successful evolutionary adaptation to an extraterrestrial environment.

In a recent book review, John G. Burke made this comment:*

> [The author] has, I believe, fallen victim on the one hand to the visionary rhetoric of the futurists, and on the other to the persuasive arguments of the technological determinists. Central to the position of both camps is the conviction that the present societal commitment to science and technology will continue into the unforeseeable future. Such a belief is hardly justified if we heed the increasingly well-defined message of the younger generation at all, not to speak of the rebellious noises currently being voiced by scientists, engineers, and political luminaries. It is far more likely that the "existential revolution" ... will be one involving a reorientation of values and institutions toward realizing the traditional ideal of man's brotherhood than one in which our natural and financial resources will be wasted by scientists seeking to create a new species of hominid or by technologists constructing vehicles for the multi-generational exploration of space.

The transition from "man to mankind" by means of adaptation to, and eventual dependence upon, technology need not be *good*; the evolution of human space travelers into a new species of hominid inhabiting another planet is neither likely, necessary, nor good.

*John G. Burke, "Technological Man" by Victor Ferkiss (a review), American Scientist 58, 343 (1970).

SECTION THREE

Race

Introduction

The selection chosen to set the tone for this section is from *Rebellion in the Backlands,* an English translation of *Os Sertões,* a Brazilian classic. If, as I said in the introduction, the literary selections are to substitute, at least informally, for the field trips and laboratory exercises of the usual biology course, a selection from *Rebellion in the Backlands* represents a trip back into time. *Os Sertões* is written in flowery 19th century Portuguese, the flavor of which has been delicately preserved in the translation. Unlike the selections I have chosen to use elsewhere in this book, I have edited and simplified this one extensively; references to early Portuguese and Brazilian historians and other little-known personalities have been removed, Portuguese terms that require explanatory footnotes have been removed, and several lengthy digressions have also been eliminated. A short glossary of regional terms follows the selection.

One appeal of Euclides da Cunha's account of the backwoodsmen (*jagunços*) of northeastern Brazil lies in the detail with which he describes their origins. Before the arrival of the Portuguese in South America, there were, to be sure, numerous Indian tribes living throughout what is now Brazil. Those persons now known as Brazilians (that is, the 19th century Brazilians of whom Cunha is writing) are a complex mixture of pre-Colombian Indians, the early Portuguese immigrants (a disproportionate number of whom were restless males), and Negro slaves brought from Africa. These are the human ingredients with which Cunha must work. The time interval within which the characteristics of the Brazilians were molded, unlike that in the case of Europeans for example, is short and rather well defined. The story that Cunha tells, therefore, is not an impossibly complicated one.

Nevertheless, its intricacies serve just the same to reveal the extent to which the geneticist simplifies racial matters. In his effort to reduce (as I shall be doing in the essays that follow) the concept of "race" and of "racial differences" to terms that are definable, that are not confounded with other terms, and that encompass entire families, entire populations of individuals, and even collections of populations, the geneticist expresses his thoughts in phrases that are almost of mathematical simplicity. Needless complexities imposed by living, breathing things are cast away so that only the stark fundamentals remain. Cunha, in contrast, has no fear of complexity. He paints with words — at one time using broad strokes to orient his reader within a vast country, at another supplying the minute details that radiate the parched heat of the *caatinga* or the smell of horses and cattle.

The 19th century notion of race encompassed a hodgepodge of human characteristics, some of which are cultural in origin, some biological, some environmental, and still others due to the interaction of any combination of culture, biology, and environment. Racial concepts are scarcely less confused in the minds of many persons today. Euclides da Cunha can be excused, therefore, if he thought that national boundaries eventually delineate races, or if he refers to a cowboy's characteristic ambling gait as a racial characteristic. Much of his information on racial matters Cunha got from his biological and anthropological contemporaries (where else could he turn?), and yet he remained skeptical of much of what they told him. This is the dilemma that confronts anyone who seeks information outside his own experiences. Does one believe the experts? Or, if a person distrusts the advice that experts have offered, is he to reject it while relying on his own intuition? Darwin accepted prevailing ideas on biological inheritance even though they confronted him with insurmountable difficulties. Cunha dutifully cites the contemporary experts on racial matters but then expresses his own distrust of their views.

In order to diverge either culturally or biologically, two populations must be separated (geographically for most species; culturally in the case of human populations living in the same geographic area) one from the other; to interbreed, members of the two populations must find one another. It is the ebb and flow of humanity in Brazil that Cunha describes so well. The river currents, the gentle and frequently inundated watershed of the state of Minas Gerais, Brazil's rocky coastal escarpment, the political role of the Church, the feudal system of land ownership, trade barriers within Brazil, and many other details influenced the eventual convergence of Paulistas (southerners) and Baians (northerners) at Canudos, the site of Brazil's backland rebellion, and the nature of the population that arose there. In the case of human beings, these are the sorts of elements that act in concert to determine what the geneticist refers to briefly as the isolation of populations or the migration between them. At one extreme in Brazil there was the Vatican exerting its influence on European governments to change the course of events in South America; at the other extreme there was the individual's inability to paddle a canoe upstream for long against a too-swift current.

Still another consideration was instrumental in the choice of *Rebellion in the Backlands* for this section. It is a striking account of guerrilla warfare, of the tenacity with which a fierce and free people will defend their home (three persons – an old man, an old woman, and a small child – were given safe conduct through the army's lines during the final battle), and the problems that confront a standard army when it embarks upon guerrilla-type operations. The final battle for Canudos, the dingy city of mud huts in Brazil's desolate *sertão*, has been compared with the Battle of Stalingrad; today, a more apt comparison would be the battle for Hue during the 1968 Tet offensive. Anyone interested in the lingering war in Vietnam would do well to read Euclides de Cunha's *Rebellion in the Backlands*.

Rebellion in the Backlands*

Euclides da Cunha

It is our belief that [confusion over Brazilian peoples] happened for the reason that the essential scope of the investigations had been reduced to the search for a single ethnic type, when, the truth is, there are many of them. We do not possess unity of race, and it is possible we shall never possess it. We are predestined to form a historic race in the future, providing the autonomy of our national life endures long enough to permit it. In this respect we are inverting the natural order of events. Our biological evolution demands the guaranty of social evolution.

We are condemned to civilization. Either we shall progress or we shall perish. So much is certain, and our choice is clear.

This is scarcely suggested, it may be, by the heterogeneity of our ancestral elements, but they are reinforced by another element equally ponderable: a physical milieu that is wide and varied and, added to this, varied historical situations which in large part flow from that milieu. This is a subject which it will be necessary to consider for a few moments.

Brazil is divided into three clearly distinct zones: the definitely tropical zone, which extends through the northern states to southern Baía, with an average temperature of 88.8 degrees; the temperate zone, extending from São Paulo to Rio Grande, by way of Paraná and Santa Catarina, between the fifteenth and twentieth isotherms; and, as a transition, the subtropical zone, extending over the north central portions of certain states, from Minas to Paraná.

Here, obviously, are three distinct habitats. Even within their more or less definite limits, however, there are circumstances which diversify them. We shall indicate these in a few rapid strokes.

The disposition of the mountains in Brazil, great upheaved masses which follow the coast in a line perpendicular to the southeast, determines the primary distinctions over large tracts of territory which lie to the east, creating a significant climatological anomaly. The fact of the matter is, the climate here, entirely subordinated to geography, violates the general laws that ordinarily govern it. Starting from the tropics, on the side of Ecuador, its astronomical

*Reprinted from Rebellion in the Backlands by Euclides da Cunha (trans. Samuel Putnam) by permission of the publisher, The University of Chicago Press. Copyright 1944. Copyright under International Copyright Union by the University of Chicago.

determination by latitudes yields to disturbing secondary causes, and it is, abnormally, defined by longitude.

This is a well-known fact. In the extensive strip of coast which runs from Baía to Paraíba more marked changes may be observed accompanying the parallel to the east then along the meridian northward. Those differences in climate and in natural features which in the latter direction are imperceptible stand out clearly in the former. All the way to the far northern regions, Nature exhibits the same unvarying exuberance in the great forests that border the coast; so that a stranger at a rapid glance would believe this to be a most fertile tract, of wide extent. On the other hand, beginning with the thirteenth parallel, the forests conceal a vast strip of sterile land, a barren tract, displaying all the inclemencies of a region in which the thermometric and hygrometric readings, marked by exaggerated extremes, vary in inverse ratio.

This is revealed by a brief journey to the west, starting from any point along the coast. The charm is broken, the beautiful illusion gone. Nature is here impoverished: no more great forests and mountaintops, but deserts and depressions, as the region is transformed into the parched and barbarous backlands with their intermittent streams and endless stretch of barren plains, forming a huge dais for the woebegone landscapes of the drought.

The contrast is most striking. A little more than a hundred miles distant are regions the exact opposite of this, with conditions of life that are equally different. It is as if one had found one's self suddenly in a desert.

And, certainly, those waves of humanity which, during the first two centuries of settlement, swept over the northern tracts on their way west to the interior must have encountered obstacles more serious than the roll of seas and mountains, when they came to cross the meager, barestripped caatingas. The failure of the Baians to penetrate to the interior of the country — and they, by the way, preceded the Paulistas — is an eloquent case in point.

The same, however, is not true of the tropics to the south. Here, the geological warp of the earth, matrix of its interesting morphogeny, remains unalterable over large expanses of the interior, creating the same favorable conditions, the same flora, a climate greatly improved by altitude, and the same animated appearance as far as natural features are concerned.

The huge bulwark of the granite cordillera, standing perpendicular to the sea, along its inner slopes falls away gently in vast, rolling plains. It forms the abrupt, steeply inclined scarp of the plateaus.

Upon the plateaus, the landscape pictures are more ample and luxuriant, without the exaggerated overwhelming aspect of the mountains. The earth is manageable, while the climate, moderately warm, rivals in mildness the admirable one of southern Europe. It is not here, as in the north, exclusively governed by the southeast wind. Blowing down off the high plains of the interior, the northwest wind is the dominant factor, as throughout the whole of that extensive zone which ranges from the elevated lands of Minas and Rio to Paraná, by way of São Paulo.

We have done little more than sketch in these broad divisions, but enough has been said to show that there is an essential difference between the south and the north, two regions that are absolutely distinct as regards meteorological conditions, the lay of the land, and the varying transitions between the inland and the coast. Coming down to a closer analysis, we shall discover special aspects that are still more to the point. We shall avoid going into an extended explanation of the subject and shall confine ourselves to the most significant examples.

We have seen, in the preceding pages, that the southeast wind is the predominant regulator of the climate along the eastern seaboard but that it is replaced in the southern states by the northwest wind and in the far north by the northeaster. But these winds in their turn disappear in the heart of the plateaus before the southwest wind, which, rushes down on Mato Grosso, occasioning thermometric variations that are out of all proportion, adding to the instability of the mainland climate, and subjecting the central regions to an extremely harsh set of conditions, differing from those that we have rapidly outlined above.

Indeed, it may be said that Nature in Mato Grosso lives up to reported exaggerations. It is quite exceptional, unique; there is nothing like it anywhere. All the wild grandeur, all the inconceivable exuberance, along with a maximum of brutality on the part of the elements — qualities which the eminent thinker, in a hasty generalization, ascribed to Brazil as a whole — exist here in reality and are manifested in astounding landscapes. Beholding these landscapes, even with the cool eye of the naturalist who is not given to rhetorical descriptions, one realizes that this anomalous climate is one that affords the most significant example of the wide variations of environment to be found in Brazil.

There is nothing like it, when it comes to a play of antitheses. The general aspect of the region is one of extreme benignity — the earth in love with life; fecund Nature in a triumphant apotheosis of bright, calm days; the soil blossoming with a fantastic vegetation — fertile, irrigated with rivers that spread out to the four corners of the compass. But this opulent placidity conceals, paradoxically, the germs of cataclysms, which, bursting forth always with an unalterable rhythm, in the summertime, heralded by the same infallible omens, here descend with the irresistible finality of a natural law. It is difficult to describe them, but we shall endeavor to give a sketch.

When the hot, moist squalls have blown for some days from the northeast, the atmosphere becomes motionless, stagnant — not a breath of air stirring. Then it is that "Nature, as it were, droops in an ecstasy of fear; not even the tops of the trees are swaying; the forests, frightfully still, appear to be solid objects; the birds cower in their nests and sing no more."

When, however, one looks up at the sky, there is not a cloud in sight! The limpid-arching blue is lighted still by a sun that is darkened, as if in eclipse. The atmospheric pressure, meanwhile, slowly but constantly drops, stifling all life as it does so. For moments, a dense cloud cumulus with copper-colored borders looms darkly on the southern horizon. Then comes a breeze whose velocity rapidly increases, turning into a high wind. The temperature falls in a few minutes, and, a moment later, the earth is shaken by a violent hurricane. Lightning flashes; thunderbolts resound in a sky that is lowering now; and a torrential rain descends on the vast expanse of plains, wiping out in a single inundation the uncertain watershed that crosses them, uniting the sources of the rivers and embroiling their beds in a limitless overflow.

The assault is a sudden one. The cataclysm bursts precipitately in the vibrant spiral of a cyclone. Houses are unroofed; the aged carandá trees are bent double, moan and split; the hills are islands; the plains deep in water.

An hour later the sun is shining triumphantly in the purest of skies! The restless birds are singing in the dripping foliage; the air is filled with gentle breezes — and man, leaving the shelter to which he had tremulously repaired, comes forth to view, amid the universal revival of Nature, the damage wrought by the storm. Trunks and boughs of trees rent by lightning and twisted by the winds; cottages in ruins, their roofs strewn over the ground; muddy rivers overflowing their banks with the last of the downpour; the grass of the fields beaten down, as if a herd of buffaloes had passed that way —sorry reminders, all, of the tempest and its fulminating onslaught.

Some days later the winds once more begin blowing up slowly from the east; the temperature begins to mount again; the barometer drops, little by little; and the feeling of general uneasiness constantly increases. This keeps up until the motionless air is caught in the formidable grip of the pampeiro, and the destructive tempest arrives, blowing in turbulent-whirling eddies, against the same lugubrious background, reviving the same old cycle, the same vicious circle of catastrophes.

And now, proceeding northward, we encounter, in contrast to such manifestations as these, the climate of Pará. Brazilians of other latitudes do not know a great deal about this climate. Mildly warm mornings, of about 73 degrees in temperature, coming unexpectedly after rainy nights; dawns that are a glowing revelation, bringing unlooked-for metamorphoses: trees which the evening before were bare now decked with flowers; marshy bogs changed into meadows. And then, again, complete transformations: silent forests, half-naked boughs with their parched or withered leaves; an air that is still and lifeless; branches shorn of their recently opened buds, whose dead, dried petals fall upon an earth that lies motionless, enervated, in a heat of 95 degrees in the shade. "On the following morning, the sun rises without clouds, and in this manner is completed the cycle — spring, winter, and autumn on a single tropical day."

The constancy of the climate is such that one is not aware of the passing of the seasons, which, all the while, as on a dial, follow one another in the hours of one day, although the average daily temperature throughout the entire year shows a variation of only one degree or a little more. Thus does life strike a balance, with a consistency that is not to be perturbed.

On the other hand, in the Upper Amazon region to the west, we come upon a new habitat, with different characteristics. And this, needless to say, imposes a painful process of acclimatization on the inhabitants of the bordering territories.

Here, at the height of the summer heat, when the last gusts of wind from the east are dying on the heavy-laden air, the thermometer comes to take the place of the hygrometer in the study of climate. The grievous round of human existence in these parts is dependent upon the emptying and filling of the great rivers. These streams rise in a most extraordinary manner. The Amazon, with a tremendous leap from its bed, in the course of a few days raises the water level to something like fifty-six feet. It expands in vast overflows, in a highly complicated network of pools and channels forming an inland sea shot by strong currents, from which emerge the green-covered islanded jungle pools.

The filling of the river brings a stoppage of life. Caught in a network of creeks, man displays a rare stoicism, in the presence of a fatality which he is powerless to avert, and awaits the end of this paradoxical winter with its high temperatures. Summer is the ebb season. The inhabitants then resume their rudimentary activities, of the only sort compatible with a natural environment that displays so wide a range of manifestations, rendering impossible the exertion of any sustained effort.

Such an environment encourages a frank parasitism. Man imbibes the milk of life by sucking on the tumid chalices of the siphonias.

But in this singular yet typical climate there are other anomalies that make it worse. The alternation of rainy and dry season, coming like the systole and diastole of one of earth's major arteries, is not enough. There are other factors which render futile for the one not born in these parts any attempt at a real acclimatization. Many times in the season when the rivers are filling, in April or in May, in the course of a calm, clear day, the air will be suddenly chilled with cold squalls from the south. It is like an icy blast from the pole. The thermometer then drops suddenly, many degrees, and for a number of days an unnatural situation prevails.

The ambitious pack-peddlers who, spurred on by hope of gain, have ventured into these parts, and the Indians themselves, who are hardened to the climate by adaptation, now take shelter in their insubstantial huts, huddling around wood fires. All work and other activities once more cease. The great solitudes, now under water, are depopulated; the fish die of cold in the rivers; the birds in the silent forests either die or take flight; nests are empty, and even the wild beasts vanish from sight, seeking refuge in the deepest holes they can find. In brief, all this marvelous wealth of equatorial nature, fashioned by the brilliant tropical suns, now presents the cruelly desolate and mournful appearance of a region at the poles. It is the cold season, or the time of friagem.

However, we shall have to bring this hasty sketch to a close. The northern backlands, as we have already seen, have in turn new climates to show, with yet other biological exigencies. Here, the same alternation of fair seasons and foul is, perhaps, more harshly reflected under other forms. Accordingly, if we stop to consider that these various climatic aspects do not represent exceptional cases but make their appearance, all of them, from the tempests of Mato Grosso to the cyclones of the northern drought area, with the periodicity which is immanent in inviolable natural laws, we shall then have to agree that the variability of our physical milieu is complete.

Reflection of Environment in History

Considering Brazilian history from a general point of view, leaving out of account for the moment the disturbing effect of nontypical details, we may behold diversified situations already taking shape in our colonial period. With the land under feudal rule, divided up among the fortunate proprietors on whom it had been bestowed, the settlement of the country was undertaken from north and south, with the same identical elements, and with the same show of indifference in either case on the part of a metropolis [Portugal] whose gaze was still turned to the last mirages of the "marvelous Indies"; and it was then that the radical separation between south and north began.

There is no need here for us to go into the decisive factors in the case of the two regions. They are two distinct histories, registering movements and tendencies opposed to each other. Two societies in process of formation, alienated by their rival destinies, one wholly indifferent to the mode of life of the other, and both all the while evolving under the influence of a single administration. In the south new tendencies were developing, a greater division of labor, more vigor in a stock that was hardier and more heterogeneous, more practical, and adventurous — a broad progressive movement, in short. And all this stood in contrast to the agitated, at times more brilliant, but always less productive life of the north: scattered and disunited captaincies, yoked to the same routine, amorphous and static, fulfilling the limited round of functions incumbent upon the pensioners of a distant court.

History here is more theatrical, less eloquent. There arise heroes, but they are great of stature only by contrast with the common run of men about them: brilliant, vibrant pages of history, but cut short, without definite objective, with the three formative races, wholly divorced from one another, playing their part in it all.

Even in the culminating period, that of the struggle with the Dutch, Henrique Dias' Negroes, Camarão's Indians, and Vieira's Lusitanians remained distinctly separate in their army tents. If they were separated in war, the distance between them grew in time of peace. The drama of Palmares [a 17th century runaway-slave colony] , the incursions of the Indians, the conflicts on the border of the backlands — all violated that truce which had been struck against the Dutch.

With the coastal region prisoned between the inaccessible backlands and the sea, there has been a tendency to hand down the old colonial alignment to our day, thanks to an obstinate and stupid centralization which has achieved the anomaly of dislocating in a new land the moral ambient of an old society. Fortunately, there was the impact of the impetuous wave from the south.

In the latter region, where acclimatization was a speedier process, in an environment that was less adverse, the newcomers soon attained an unwonted vigor. Out of the absorption of the aboriginal tribes came the mixed descendants of the backland conquests, the daring mamelucos. The Paulista — and this name, in its historic signification, takes in the sons of Rio de Janeiro, Minas, São Paulo, and regions south — now arose as an autonomous type, adventurous, rebellious, freedom-loving, the perfect model of a lord of the earth, freeing himself, rebel-like, from a distant rule, leaving behind him the sea and the galleons of the metropolis, casting in his lot with the hinterland, and carving out the unsung epic of the "Bandeiras."

All this admirably reflects the influence of environmental conditions. There was no racial distinction whatsoever between the colonizers from the south and those from the north. In either case, there was a prevalence of those same elements which were the despair of Diogo Coelho: "Worse here in the land than pestilence. . . . " But in the south the vital force of temperament remaining with those who had bested the unyielding ocean was not wasted by an enervating climate but found a fresh component in the very strength of earth, which changed it, and for the better. Man here felt himself strong and capable. With the theater of his great crimes shifted but a little, he might bring to bear upon the impervious backlands the same audacity which had hurled him into African circumnavigations.

In addition — and we must touch upon this point at the risk of scandalizing our puny historiographers — the mountains freed him from the preoccupation of having to defend a littoral where the bark of the covetous foreigner might put in. The Serra do Mar, towering perpendicularly above the Atlantic, like the curtain of a huge bastion, has played a notable role in our history. Against its cliffs the warrior passion of the Cavendishes and the Fentons beat in vain. On its heights, letting his gaze roam round about him over the plains, the newcomer felt secure. He felt himself on battlements that were not to be moved, which served at once as his protection against the foreign invader and against the cavalier from the metropolis. The mountains, arched like hoops of stone about a continent, were an isolating factor in both the ethnic and the historical sense of the word. They canceled out that irrepressible attraction to the seaboard which existed in the north, a seaboard reduced to a narrow strip of swamps and reefs, which held out nothing to excite man's covetousness; while there arose, superior to all fleets, intangible in the depths of the forests, the mysterious lure of the mines.

What is more, the special topographical relief of these mountains turns them into a condenser of the first order, in precipitating the evaporated moisture of the ocean. The rivers which flow down the slopes may be said in a manner of speaking to rise in the sea, rolling their waters in an opposite direction from the coast, carrying them to the interior, making straight for the backlands. They inevitably put the newcomer in mind of exploring expeditions. The earth attracts man, calls him to her fertile bosom, enchants him with her great beauty, and ends by snatching him, irresistibly, into the river currents.

The Tieté in its course is an eloquent case in point, giving direction to this conquest of the land. Whereas on the São Francisco, the Parnaíba, the Amazon, and all the other streams of the eastern seaboard, in proceeding up them to the interior, one has to battle the currents and the cataracts which fall from the terraces of the plateaus, the Tieté without the exertion of rowing carries the inhabitants of the region to the Rio Grande and thence to the Paraná and the Parnaíba. It was the means of penetration into Minas, into Goiaz, Santa Catarina, the Rio Grande do Sul, Mato Grosso, all Brazil. Following this line of least resistance, which at the same time affords the clearest outline of our colonial expansion, the bandeiras did not find, as in the north, a sterile earth and the intangible barrier of ugly deserts to slow their progress.

It is easy to show how this distinction of a physical nature sheds light on the anomalies and contrasts to be found in the course of events in the two parts of the country, above all in the acute period of colonial crisis in the seventeenth century. The Dutch rule at this time was centered in Pernambuco, but it affected the entire east coast from Baía to Maranhão; and in the struggle that ensued there were certain memorable encounters in which our three formative races joined forces against the common enemy. The man of the south, on the other hand, held himself absolutely aloof from this movement and showed, by rebelling against the decrees from the metropolis, how completely divorced he was from his battling fellow-countrymen. He was in a way an enemy fully as dangerous as the Dutch. A strange race of mestizos, given to uprisings, swayed by other tendencies, guided by another star, setting off in other directions, these southerners resolutely trod underfoot all restraining bulls and letters patent. In stubborn reaction to the Jesuits, they entered into an open struggle with the Portuguese court. As for the Jesuit fathers, they, forgetting the Dutch, had recourse to Madrid through Ruy de Montoya, and to Rome through Dias Taño; which showed that they looked upon the southerner as being the more serious foe.

Indeed, while in Pernambuco Von Schoppe's troops were setting up the Nassau government, in São Paulo the groundwork was being laid for the somber drama of Guaira. And when the restoration in Portugal led to the defeat of the invader all along the line, bringing the exhausted combatants together once more, the men of the south made it still more plain that theirs was a separate destiny by taking advantage of this very fact to assert their full and free autonomy in the one-minute rule of Amador Bueno.

There is no greater contrast in all our history. It is one truly national in significance, but one of which there is barely a glimpse to be had in the spectacular courts of the governors of Baía, where the Society of Jesus held sway, with the privilege of the conquest of souls, a casuistic euphemism cloaking a monopoly of the native's good right arm.

In the middle of the seventeenth century the contrast becomes still more pronounced. The men of the south were spreading out over the interior, making their way as far as the extreme limits of Ecuador. Down to the middle of the eighteenth century, the settlers followed the confused trails of the bandeiras. Wave after wave of them came, with the untiring fatality of a natural law; and, indeed, they did represent a vast potential, these great warrior caravans, these human waves let loose on the four corners of the compass, stamping over their country at every point, discovering it after the discovery, laying bare the gleaming bosom of its mines.

Leaving behind them the seaboard, which reflected the decadence of the metropolis, along with all the vices of a nationality in process of hopeless decomposition, these pioneers, making their way to the far lands of Pernambuco and the Amazon, gave the impression of being of another race, in their intrepid daring and their ability to withstand adversity. When the incursions of the savage threatened Baía, or Pernambuco, or Paraíba, and the runaway-slave quilombos, last refuge of the rebellious African, were set up in the forest, the southerner, as the crude epic of Palmares tells us, thereupon arose as the classic hero vanquishing all perils, the chosen one for the undertaking of giant hecatombs.

The truth is, the northerner did not possess a physical environment which endowed him with an equal amount of energy. Had such been the case, the bandeiras from the east and those from the north would have come together and crushed the native, who would have disappeared without leaving any trace. But the northern colonist, to the west and south, found himself face to face with hostile Nature and speedily turned back to the coast; his was not the daring of those conquerors who feel themselves at home in a friendly land; his was not the self-assurance inspired by the very attractiveness of luxuriant and easily accessible regions. The explorations which were here begun in the second half of the sixteenth century are but a pallid imitation of the irruptions of the bandeirantes of São Paulo.

Caught between the coastal canebrakes and the backlands, between the sea and the desert, in a blockade that was rendered more formidable by the action of the climate, the northerner wholly lost that upstanding spirit of rebellion which resounds so eloquently throughout all the pages of southern history. Such a contrast, certainly, is not based on primordial racial factors.

Having thus outlined the environmental influence on our history, let us see now what its influence has been upon our ethnic formation.

Action of Environment in the
Initial Phase of the Formation of Races

Let us return to our point of departure. Agreed that environment does not form races, in our own special case, in various parts of the country, it produces an excessive variation in the degree of admixture of the three essential elements. Through the very diversity of the conditions of adaptation, it prepares the way for the appearance of different subraces. But let us confine ourselves to our subject by outlining rapidly the historical antecedents of the jagunço. [Jagunço is used as if it were synonymous with sertanejo, a "hillbilly" of northeastern Brazil — of the region known as the sertão. Just as in America the word "cowboy" is used in referring to many Westerners other than cattlemen, so the word vaqueiro is used as a general term for jagunços and sertanejos. In a similar way, gaucho — originally a term for cowboys of southern Brazil — is now used in referring to all persons from that area.]

As we have already seen, Brazilian racial formation in the north is very different from that in the south. Historic circumstances, deriving in large part from physical ones, gave rise to initial diversities in the intermingling of the races — diversities which have been handed down to our own day. The march of settlement from Maranhão to Baía reveals these differences.

The First Settlers

It was a slow process. The Portuguese did not approach the northern seaboard with that vital strength which comes from dense migrations, great masses of invaders capable of preserving, even when uprooted from their native soil, all those qualities acquired in the course of a long historical apprenticeship. They were scattered, parceled out in small bands of condemned exiles or counterfeit colonists, lacking in the virile mien of conquerors. They still were dazzled by visions of the Orient. Brazil was the land of exile, a huge garrison for the intimidating of heretics and backsliders, all those victims of the somber let-him-die-for-it justice of those days. And so it was, in these early times, the reduced number of settlers stood in contrast to the vast expanse of the land and the size of the native population. The instructions given, in 1615, to Captain Fragoso de Albuquerque, concerning the regularizing with the Spanish ambassador in France of the truce agreement with La Revardière, are sufficiently clear in this respect. Here it is stated that "the lands of Brazil are not unpopulated, for the reason that there exist in them more than three thousand Portuguese."

This for the whole of Brazil — more than a hundred years after the discovery.

As Ayres de Casal tells us, "the population grew so slowly that at the time of the decease of Sr. D. Sebastião (1580), there was not a settlement outside the island of Itamaracá, the inhabitants of which locality numbered some two hundred, with three sugar plantations."

When, a few years later, Baía was a little better populated, the unfavorable disproportion between the European and the two other elements still continued, in perfect arithmetic progression. According to Fernão Cardim, there were two thousand whites, four thousand Negroes, and six thousand Indians. It is obvious that the native element was for a long time the predominant one. In the first intermarriages, therefore, it must have had a large influence.

The newcomers who descended upon these regions were, moreover, of a type adapted to a large-scale miscegenation. Men of war, without homes, given to the solitary life of camps, or else exiles and corrupt adventurers. Concubinage with the Indian women degenerated into open debauchery, from which not even the clergy were free. Padre Nobrega clearly establishes this fact, in his celebrated letter to the King (1549), in which, depicting with an ingenuous realism the laxity of manners, he declares that the interior of the country is full of the offspring of Christians, who increase and multiply in accordance with heathen custom. The first intensive mixture of races, between the European and the Indian, occurred, then, in these early times. "Soon," says Casal, "the Tupinquins, who are pagans of good disposition, were christened and married off to Europeans, there being innumerable natural whites in the country with the blood of the Tupinquins in their veins."

On the other hand, although they existed in large numbers within the bounds of the kingdom, the Africans in this first century played a minor role. In many places there were few of them to be found. They were few, according to the trustworthy narrator just quoted, in Rio Grande do Norte, "where the Indians were long since reduced, despite their ferocity, and where their descendants through alliances with Europeans and Africans have augmented the classes of whites and brown-skins."

These excerpts are significant. Without any preconceived ideas on the subject, it may be asserted that the disappearance of the native in the north was due not so much to actual extermination as to repeated intermarriages.

We know, moreover, that the landed proprietors were anxious to make the most of such alliances in capturing the affections of the native. This attitude reflects the instincts of the metropolis, as is shown in the various royal charters from 1570 to 1758, in which, despite a never interrupted train of hesitations and contradictions, there appeared a desire to mitigate the lust for gain on the part of the colonists, bent upon the enslavement of the savage. Some of these charters, that of 1680, for example, extended the royal protection to the point of decreeing that lands shall be granted the heathen, "even those uncultivated lands already given to others, inasmuch as those same Indians ought to have the preference, being the natural lords of the land."

In this persistent attempt at the assimilation of the native, an important part was played by the Society of Jesus, which, compelled in the south to submit to compromises, ruled supreme in the north. Putting aside any unworthy intentions, the Jesuits in this region performed an ennobling task. At the least, they were rivals of the colonist, who was bent on gain. In the stupid clash of perversity with barbarism, these eternal exiles found a function that was worthy of them. They did much. They were the only disciplined men of their day. Although the attempt to elevate the mental state of the aborigine to the abstractions of monotheism may have been a chimerical one, it had the effect of attracting him for a long time into the stream of our history.

The record of the missions in the north, throughout the tract from Maranhão to Baía, reveals a slow effort at penetration into the heart of the backlands completing in a manner the feverish activity of the bandeiras. If these latter, despite the disturbances which they brought with them, diffused widely over the newly discovered regions the blood of the three races, thereby occasioning a general miscegenation, the organized settlements, on the other hand, centers of the attractive force exerted by the apostolate, had the effect of unifying and integrating the native tribes, fusing their small communities into villages. Penetrating, as the result of a century-long effort, deep into the backlands, the missionaries were responsible for saving in part this factor in our racial picture. Some historians, distracted by the coming of the African on a large scale, an event which, beginning at the end of the sixteenth century, continued uninterruptedly down to our time (1850), have felt that the latter was the best ally of the Portuguese in the colonial epoch and, accordingly, attribute to him generally an exaggerated influence upon the formation of the inhabitant of the northern backlands. However, it is debatable as to whether they exercised any profound influence on the back-country regions.

It is certain that the Afro-Lusitanian association is an old one, antedating even the discovery, since it goes back to the fifteenth century, to the azenegues and jalofos of Gil Eannes and Antão Gonçalves. In 1530 there were more than ten thousand Negroes in the streets of Lisbon, and the same was true of other places. In Evora the Negroes were in a majority over the whites. The words of a contemporary are a document here: "We see so many captives brought into the kingdom and their numbers so increasing that, if this keeps up, as I see it, they will become the natives and will outnumber us."

The Origin of the Mulatto

Thus, the origin of the mulatto is to be looked for outside our own country. The first intermarrying with the African occurred in Portugal. With us, naturally, the number of such marriages tended to increase. However, the dominated race found its capacities for development annulled by the social situation that prevailed. A powerful organism, given to an extreme humility, without the Indian's rebelliousness,* the Negro at once had the whole burden of colonial life placed on his shoulders. He was a beast of burden, condemned to labor unceasingly. Old ordinances, setting forth "how one may rid himself of slaves and beasts upon finding them sick or maimed," show the brutality of the age. Moreover, and the point is incontrovertible, the numerous slaves imported were concentrated on the seaboard. The great black border hemmed the coast from Baia to Maranhão but did not extend far into the interior. Even in open revolt, the humble Negro, become a dread fugitive, appeared to avoid the heart of the country. Palmares, with its thirty thousand runaway slaves was, after all, only a few miles distant from the coast.

On the seaboard the fertility of the earth tended to retain there two of the three elements, while freeing the Indian. The extensive culture of sugar cane, imported from Madeira, led to the backlands' being forgotten. Already before the Dutch invasion, there were a hundred and sixty sugar plantations from Rio Grande do Norte to Baía. And this exploitation, on an expanded scale, was later to increase in a rapid crescendo.

The African element, in any event, remained in the vast canebrakes of the coast, fettered to the earth and giving rise to a racial admixture quite different from the one taking place in the recesses of the captaincies. In the latter place the free Indian roamed, unadapted to toil and always rebellious or barely held in check in the settlements through the tenacity of the missionaries. Negro slavery, the result of the colonist's self-interest, left the padres less encumbered with manual labor than in the south and with more time for catechizing. As for the pioneers themselves, upon arriving at this last stage of their daring journey, they had all their combativeness extinguished. Some of them brought an adventurous life to a close by seeking the profit to be made on the stockbreeding farms which had been opened on the great estates.

*It may be pointed out that present-day anthropologists and historians are not inclined to agree with this view. Arthur Ramos, for example, has this to say: "It has been repeatedly but erroneously asserted, by Brazilian historians and sociologists, that the Negro in Brazil ... was a passive element, resigned to a state of slavery. According to these historians, the African Negro, a humble, docile creature, permitted himself to be taken, and submitted without protest to slave labor. This is a view which is categorically refuted by a study of history and sociology. Such a study shows us, on the contrary, that the Negro was never this docile, submissive type, incapable of reacting to conditions ... He was a good worker but a bad slave. Several centuries of slavery and slave revolts, not only in Brazil, but in other parts of the Americas as well, show us what his true reactions were. Those reactions ranged all the way from flight to suicide, from individual escape to great collective uprisings." Professor Ramos stresses the fact that if the Negro was enslaved in preference to the Indian, it was because he was "better adapted than the Indian to agricultural labor, by reason of his cultural background." (Footnote from Putnam's translation.)

In this manner a thoroughgoing distinction arose between intermarriages in the backlands and those on the seaboard. As a consequence, with the white element the common denominator in both cases, the mulatto appeared as the principal result in the latter, and the curiboca in the former, instance.

Origin of the Jagunços

There can be no doubt that there is a notable trace of originality in the formation of our backlands population, we shall not say of the North, but of subtropical Brazil. Let us try to sketch it in; and, in order not to stray too far from the subject, let us keep close to that theater in which the historic drama of Canudos was unfolded, by traversing rapidly the reaches of the São Francisco, that "great highway of Brazilian civilization," as one historian has put it.

From the bird's-eye view given in the foregoing pages, we have seen that this river flows through regions that differ greatly in character. At its headwaters it spreads out, its expanded basin with its network of numerous tributaries taking in the half of Minas, in the zone of mountains and forests. Later, along its middle portion, it narrows, in the extremely beautiful region of the Campos Gerais. Along its lower course, downstream from Joazeiro, where it is confined between slopes which render its bed uneven and which twist it about in the direction of the sea, it becomes poor in tributaries, almost all of them intermittent, and flows away hemmed in a single corredeira of several hundred miles in length, extending to Paulo Affonso — and here it is that it cuts through the semidesert tract of caatingas.

In the threefold aspects of this river we have a diagram of the course of our history, one that reflects in parallel fashion its varying manifestations. It balances the influence of the Tieté. Whereas this latter river, with a course incomparably better suited to purposes of colonization, became the chosen pathway of pioneers seeking, above all, the enslavement and corruption of the savage, the São Francisco at its headwaters was essentially the center of the movement to the mines; in its lower course it was the scene of missionary activity; and, in its middle region, it became the classic land of the herdsman, representing the only mode of life compatible with existing social and economic conditions in the colony. Its banks were trod alike by the bandeirante, the Jesuit, and the vaqueiro.

When, at some future date, a more copious supply of documents shall enable us to reconstruct the life of the colony from the seventeenth century to the end of the eighteenth, it is possible that the vaqueiro, wholly forgotten today, will stand out with that prominence which he deserves, by reason of his formative influence on the life of our people. Brave and fearless as the bandeirante, as resigned and tenacious as the Jesuit, he had the advantage of a supplementary attribute which both the others lacked — he had his roots fixed in the soil.

As for the bandeiras, there were two sides to their activity, which were at times distinct, at other times mingled in confusion. Sometimes they fell upon the land and sometimes upon man. Sometimes they went in search of gold and sometimes in search of slaves. But one thing they did do was to discover huge tracts of land — land which they did not cultivate and which, it might be, they left more of a desert than it was before, as they passed rapidly on through the villages and the abandoned mine clearings. Their history, at times as undecipherable as the deliberately obscure entries in their logbooks, was a successive alternation and combination of these two stimuli, depending upon the personal temperament of the adventurers concerned or the greater or less degree of practicability of the enterprises planned. In the permanent oscillation between these two motives, their really useful function, which lay in discovering the unknown, appears as an obligatory incident, an inevitable consequence to which they gave no thought.

It was during this brief period when, to all appearances, nothing of note was occurring on the seaboard beyond the struggle with the Dutch and, in the heart of the plateau country, the extraordinary surge of the bandeiras, that there began along the middle reaches of the São Francisco a process of settlement the results of which were only to become apparent later.

Historical Function of the São Francisco

It was a slow process. A determining factor in the beginning was the expeditions to the mines of Moreya, expeditions which, though their fame was scant, would seem to have extended all the way to the vicegerency of Lancastro, thus opening a path for successive swarms of settlers, to the highlands of Macaúbas, beyond Paramirim. Deprived of direct routes perpendicular to the coast, which would have been shorter, but which were cut off by thick mountain walls and blocked by forests, explorers found access to the backlands by way of the São Francisco. Affording them two entrances, one at its source and the other at its mouth, and bringing the southerners to meet the men of the north, the great river from the beginning took on the aspect of an ethnic unifier, an extensive bond of union between two societies that were in ignorance of each other. Although coming from diverse points and with equally diverse backgrounds, the Paulistas, the Baians with small armies of Indian allies, or even the Portuguese — whoever they were, from wherever they hailed, the newcomers, upon reaching the heart of the backlands, seldom returned.

The land at once highly productive and readily accessible compensated them for the lost mirage of the longed-for mines. The original character of its geological structure created topographical conformations in which the highlands, the last spurs and counterforts of the maritime cordillera, were offset by vast tablelands. Its flora was complex and varied, interspersed with forests that were not so vast and impenetrable as those of the coast, along with the "charm" and "rustic beauty" of fields and plains that were suddenly lost, all of them, in the enormous glades of the caatingas. Its special hydrographic conformation, with tributaries running almost symmetrically east and west, linked it to the coast on the one side and, on the other side, to the center of the plateau region. All these were precious lures, attracting the scattered elements and leading to their fusion. And, finally, there was the opportunity for a herdsman's life, the one that was naturally suited to this region of the Campos Gerais. As it was populated, this region became a strong and autonomous one; but little was heard of it, since the chroniclers of the time paid small attention to it, and it accordingly remained utterly forgotten, not only by the distant metropolis, but by the governors and viceroys themselves. It did not produce taxes or revenues to interest a selfish-minded monarch. It represented, nonetheless, in contrast to the turbulence of the seaboard and the adventurous episode of the mines, "almost the only tranquil aspect of our culture." Aside from the rare contingents of Pernambucan and Baian settlers, the majority of the well-to-do cattlemen who sprang up here came from the south, of the same enthusiastic and energetic stock from which the bandeiras sprang.

The Jagunços: Probable Relatives of the Paulistas

There were numerous families of São Paulo which, in continuous migrations, sought out these far corners of the land; and it is to be believed, if we accept the opinion of a perspicacious historian, that "the valley of the São Francisco, long since populated by the Paulistas and their descendants, from the eighteenth century become almost an exclusive colony of them." It was, accordingly, natural that Bartholomeu Bueno, in discovering Goiaz, to his surprise should have come upon the obvious traces of predecessors, anonymous pioneers who had come there, certainly, from the east, after crossing the Paranan Range. And then, in 1697, the most notable of the cycle of gold rushes began, amid all the noise and stir created by the waves of immigrants who stamped up the eastern slopes of the Espinhaço Range, along the thalweg of the Rio das Velhas, the hardiest of all those immigrants, perhaps — passing the others by from side to side, and advancing in a contrary direction, like a reflux emanating from the north — were the bands of "Baians," a term which, like that of

"Paulista," was destined to become a generic one, being extended to take in all the northern settlers.

The Vaqueiro

There was already being formed in the middle valley of the great river a race of mestizos identical with those bold mamelucos who had sprung up in São Paulo. We shall not be indulging in too daring a hypothesis by asserting that this extraordinary type of Paulista was able to preserve intact, down to our own time, the virile and adventurous character of his ancestors. Here he remained utterly divorced from the rest of Brazil and the world, walled in on the east by the Serra Geral and shut off on the west by the broad sweep of the Campos Gerais which to this day the native believes to be boundless.

The milieu at once attracted and held him. The water gaps on one and the other side of the meridian did not invite to dispersion but rather facilitated the intermingling of the far sections of the country, binding them together in space and in time. Assuring in the interior a contiguity of the population which was in part lacking on the coast, this rude society, arising between the northeasterners struggling for the autonomy of their nascent fatherland and the southerners who were enlarging its area and who, at the same time, were supplying it with those fat herds which roamed the valley of the Rio das Velhas or which made their way down to the headwaters of the Parnaíba — this society, misunderstood and forgotten, was the vigorous core of our national life.

The first pioneers who were responsible for this creative effort, having overcome and supplanted the savage all along the line, proceeded now to capture and enslave him, making use of his capacities in the new industry which they were founding. The result was the inevitable intermarrying; and there appeared then a race of pure curibocas with almost no mixture of African blood, as is easily shown by the normal appearance of these inhabitants of the region. They sprang from the fierce embrace of victor and vanquished. They were created in a turbulent, adventurous society, settled upon a fertile land. By way of amplifying their ancestral attributes, they had a rude schooling in force and courage on those same broadsweeping campos where even today the jaguar roars with impunity and the swift-footed rhea roams — or else on the mountainsides, crumbling from surface-mining operations, when the Baians later summoned these lusty-lunged cowboys from their roundups to work in the mines. It would take too long here to trace the evolution of their character. With something of the colonist's adventurous disposition, combined with the impulsiveness of the native, they were further subjected to the influence of their environment, which, by isolating them, helped them to preserve the attributes and habits of their forebears, only slightly modified by the exigencies of their new life.

Here they stand, then, with their characteristic garb, their ancient customs, their strange adherence to the most remote traditions, their religious sentiment carried to the point of fanaticism, their exaggerated point of honor, and their exceedingly beautiful folklore and folk poetry, three centuries old.

A strong and ancient race, with well-defined and immutable characteristics, even in the major crises of life — at which times the cowboy's leather garb becomes the jagunço's flexible armor — sprung from far-converging elements, yet different from all the rest of the population of the country, this stock is undeniably a significant example of the importance of those reactions induced by environment. Spreading out through the bordering or near-by regions of Goiaz, Piauí, Maranhão, Ceará, and Pernambuco, their character became marked by a high degree of originality, which was expressed even in the houses that they built. All the villages, towns, and cities that give life to this territory show a common origin and are quite distinct in appearance from those in the north and in the south. In the south their dwellings were erected in the neighborhood of the mines or on the edge of mine clearings while in the far north they were constructed in the locality of the ancient missionary villages; but in the region here under consideration they all of them sprang up from the old cattle ranches.

We may be excused from citing examples, of which there are many to be had. Whoever views the settlements along the São Francisco, from its source to its mouth, cannot fail to observe the three types of country that we have pointed out. Leaving the Alpine-like regions, with cities perched high on the mountainsides, reflecting the incomparable daring of the bandeiras, he will cross the great campos, a huge arena made to the measure of a rude, strong, and freedom-loving people, the vaqueiros; and then, finally, he will come to the unprepossessing district laid waste by drought and elected for the slow, laborious circuits of the missionaries.

There remains for us to describe, by way of rounding out these hasty comparisons, the Jesuit foundations in this tract of territory.

Jesuit Foundations in Baiá

The truth of the matter is that the backland towns of the present day, wholly different in origin from those elsewhere, were formed from the old Indian villages which, in 1758, were wrested from the power of the padres. Confining ourselves to those that still exist today, nearest to and round about the mud-walled Canudos, we may find even within this restricted area the best examples.

Indeed, throughout the whole of this tract, which abusive concessions had placed within the power of a single family, that of Garcia d'Avila (Casa da Torre), there are some very old settlements. From "Itapicurú-de-cima" to Geremoabo, and from there, following the São Francisco, to the backland regions of Rodellas and Cabrobó, the seventeenth-century missions proceeded at a slow pace, which was to continue until our own time.

They had no historian. It is difficult today to form a picture of this extraordinary undertaking from the few documents that exist and which are too scant to permit us to trace its continuity. Those that do exist, however, are an eloquent commentary on the special case here under consideration. They inform us, in a manner that leaves no room for doubt, that, while the Negro's days were spent amid the hurly-burly of the seaboard, the native was taking up a permanent abode in villages which were to become cities. The calculating solicitude of the Jesuit and the rare self-abnegation displayed by the Capuchins and Franciscans were responsible for incorporating the native tribes into our national life; and when, at the beginning of the eighteenth century, the Paulistas stormed into Pambú and Jacobina, they gazed with surprise at these parishes which already were growing up out of the collections of tribal huts.

As for Geremoabo, we find it already a borough in 1698, which permits us to ascribe to it an origin that is a good deal more remote. Here there was a slight admixture of the native element with the African. Incomparably more spirited than is the case today, the humble village not infrequently attracted the attention of the governor-general of Brazil, principally when the rivalries of the Indian chieftains, equipped with perfectly legal letters patent from the captaincies, grew more acute. In the year 1702 the first of the Franciscan missions undertook to discipline these settlements, their efforts proving more efficacious than had the governor's threats. The tribes were reconciled, and the influx of captured Indians into the church was such that on a single day the vicar of Itapicurú baptized 3,700 catechumens.

It was, however, in the north, from the beginning of the eighteenth century, that the work of settlement was carried on most intensely, with the same racial elements. In the second half of the seventeenth century the vanguard of the southern bandeiras made its appearance in the Rodellas region. Domingos Sertão's fazenda of Sobrado was then the center of the animated life of the backlands. The effect which this rude pioneer had upon the district has not been given the attention that it deserves. With a domain situated practically at the point of convergence of the northern captaincies, being near at once to Piauí, to Ceará, to Pernambuco, and to Baía, this rural landlord made use of the restless, adventurous curibocas in the working of his half-hundred cattle ranches. Addicted to a crude variety of feudalism, which led him to transform his tributaries into vassals and the meek Tapuias into serfs, the bandeirante, once he had reached these parts and had attained his ideal of wealth and power, continued to fulfil his function of integrating the population in alliance with his humble but stubborn adversary, the priest. The northern metropolis, the fact is, unhesitatingly supported the efforts of the latter; for a long while since, the principle had been laid down of combatting the Indian with the Indian, each missionary village of catechumens being looked upon as a redoubt against the free-roaming and indomitable savage.

It was at the end of the seventeenth century that Lancastro with his catechized natives founded the settlement of Barra, against the depredations of the Acaroazes and Mocoazes; and from this point on, following the current of the São Francisco, there were to be found, one after another, a large number of missionary villages. It is, then, evident that it was precisely in this section of the Baian backlands, the one with closest ties to the other northern states — along the entire border of the Canudos region — that there developed at the beginning of our history a strong colonization movement in which the Indian played a leading role, intermingling with white man and Negro, with these latter elements never becoming so numerous as to destroy his undeniable influence.

What has happened is that this population stuck away in a corner of the backlands has remained there until now and has gone on reproducing itself without the admixture of foreign elements; it has been, as it were, insulated, and by reason of this very fact has been able to carry on the task of racial assimilation with a uniform and maximum intensity in a manner that would account for the appearance of a well-defined and well-rounded mestizo type.

Where on the seaboard there were a thousand and one complicating and disturbing factors, due to immigration and to war, while at other central points yet other impediments arose in the sweeping trail of the bandeiras, here, on the other hand, the indigenous population, in alliance with a few wandering runaway slaves, white fugitives from justice or audacious adventurers, remained the dominant one.

Causes Favorable to the Formation
of a Mestizo Race in the Backlands,
as Distinguished from Crossbreeding on the Littoral

Let us not play sophists with history. There were very powerful causes which led to the isolation and conservation of the native stock.

First of all, there were the great land grants, representing the most perduring aspect of our shamefaced feudalism. The possessors of the soil, the classic model being the heirs of Antonio Guedes de Britto, were jealous of their far-flung latifundia, which, with no boundary lines to demarcate them, made their owners the lords of the countryside, barely if at all tolerating the intervention of the metropolis itself. The erection of chapels or the establishment of parishes on their lands was always accompanied by controversies with the padres; and, although the latter won out in the end, they nevertheless, to a degree, came under the sway of these potentates, who made it difficult for new settlers or competitors to come in by turning their cattle ranches, which lay scattered around in the neighborhood of the newly formed church domains, into powerful centers of attraction for the mestizo race that inhabited the parishes.

That race accordingly developed without the influence of external elements. Devoting themselves to the herdsman's life, to which they were by nature well adapted, the curibocas or swarthy cafusos, immediate forebears of the present-day vaqueiros, being entirely cut off from the inhabitants of the south and the intensive colonization activities of the seaboard, proceeded to follow their own path of evolution, acquiring thereby a highly original physiognomy, like that of residents of another country. The royal charter of February 7, 1701, was a measure designed to increase this isolation. It prohibited, with severe penalties for infraction, any communication whatsoever between this part of the backlands and the south, the São Paulo mines. Not even commercial relations were tolerated, the simplest exchange of products being forbidden.

It is natural that the great backlands populations, like the one which grew up in the middle basin of the São Francisco, should have been formed with a preponderant admixture of Indian blood. And there they remained in banishment, evolving in a closed circle for three centuries, down to our era — completely abandoned, wholly alien to our destinies, and preserving intact the traditions of the past. Accordingly, whoever today traverses these regions will observe a notable uniformity among the inhabitants: an appearance and stature that vary but slightly from a given model, conveying the impression of an unvarying anthropologic type, one which at first glance is seen to be distinct from the proteiform mestizo of the seaboard; for, where the latter shows all varieties of coloring and remains ill defined in type, depending upon the varying predominance of the formative factors, the man of the backlands appears to have been run through one common mold, with the individuals exhibiting almost identical physical characteristics: the same complexion, ranging from the bronze hue of the mameluco to the swarthy color of the cafuso; hair straight and sleek or slightly wavy; the same athletic build; and the same moral characteristics, the same superstitions, the same vices, and the same virtues.

This uniformity, in its various aspects, is most impressive. There is no doubt about it, the backwoodsman of the north represents an ethnic subcategory that has already been formed.

The Sertanejo

The sertanejo, or man of the backlands, is above all else a strong individual. He does not have the flawless features, the graceful bearing, the correct build of the athlete. He is ugly, awkward, stooped. Hercules-Quasimodo reflects in his bearing the typical unprepossessing attributes of the weak. His unsteady, slightly swaying, sinuous gait conveys the impression of loose-jointedness. His normally downtrodden mien is aggravated by a dour look which gives him an air of depressing humility. On foot, when not walking, he is invariably to be found leaning against the first doorpost or wall that he encounters; while on horseback, if he reins in his mount to exchange a couple of words with an acquaintance, he braces himself on one stirrup and rests his weight against the saddle. When walking, even at a rapid pace, he does not go forward steadily in a straight line but reels swiftly, as if he were following the geometric outlines of the meandering backland trails. And if in the course of his walk he pauses for the most commonplace of reasons, to roll a cigarro, strike a light, or chat with a friend, he falls — "falls" is the word — into a squatting position and will remain for a long time in this unstable state of equilibrium, with the entire weight of his body suspended on his great-toes, as he sits there on his heels with a simplicity that is at once ridiculous and delightful.

He is the man who is always tired. He displays this invincible sluggishness, this muscular atony, in everything that he does: in his slowness of speech, his forced gestures, his unsteady gait, the languorous cadence of his ditties — in brief, in his constant tendency to immobility and rest.

Yet all this apparent weariness is an illusion. Nothing is more surprising than to see the sertanejo's listlessness disappear all of a sudden. In this weakened organism complete transformations are effected in a few seconds. All that is needed is some incident that demands the release of slumbering energies. The fellow is transfigured. He straightens up, becomes a new man, with new lines in his posture and bearing; his head held high now, above his massive shoulders; his gaze straightforward and unflinching. Through an instantaneous discharge of nervous energy, he at once corrects all the faults that come from the habitual relaxation of his organs; and the awkward rustic unexpectedly assumes the dominating aspect of a powerful, copper-hued Titan, an amazingly different being, capable of extraordinary feats of strength and agility.

This contrast becomes evident upon the most superficial examination. It is one that is revealed at every moment, in all the smallest details of back-country life — marked always by an impressive alternation between the extremes of impulse and prolonged periods of apathy.

It is impossible to imagine a more inelegant, ungainly horseman: no carriage, legs glued to the belly of his mount, hunched forward and swaying to the gait of the unshod, mistreated backland ponies, which are sturdy animals and remarkably swift. In this gloomy, indolent posture the lazy cowboy will ride along, over the plains, behind his slow-paced herd, almost transforming his "nag" into the lulling hammock in which he spends two-thirds of his existence. But let some giddy steer up ahead stray into the tangled scrub of the caatinga, or let one of the herd at a distance become entrammeled in the foliage, and he is at once a different being and, digging his broad-roweled spurs into the flanks of his mount, he is off like a dart and plunges at top speed into the labyrinth of jurema thickets.

Let us watch him at this barbarous steeple chase. Nothing can stop him in his onward rush. Gullies, stone heaps, brush piles, thorny thickets, or riverbanks — nothing can halt his pursuit of the straying steer, for wherever the cow goes, there the cowboy and his horse go too. Glued to his horse's back, with his knees dug into its flanks until horse and rider appear to be one, he gives the bizarre impression of a crude sort of centaur: emerging unexpectedly into a clearing, plunging into the tall weeds, leaping ditches and swamps, taking the small hills in his stride, crashing swiftly through the prickly briar patches, and galloping at full speed over the expanse of tablelands.

His robust constitution shows itself at such a moment to best advantage. It is as if the sturdy rider were lending vigor to the frail pony, sustaining it by his improvised reins of caroá fiber, suspending it by his spurs, hurling it onward — springing quickly into the stirrups, legs drawn up, knees well forward and close to the horse's side — "hot on the trail" of the wayward steer; now bending agilely to avoid a bough that threatens to brush him from the saddle; now leaping off quickly like an acrobat, clinging to his horse's mane, to avert collision with a stump sighted at the last moment; then back in the saddle again at a bound — and all the time galloping, galloping, through all obstacles, balancing in his right hand, without ever losing it once, never once dropping it in the liana thickets, the long, iron-pointed, leather-headed goad which in itself, in any other hands, would constitute a serious obstacle to progress.

But once the fracas is over and the unruly steer restored to the herd, the cowboy once more lolls back in the saddle, once more an inert and unprepossessing individual, swaying to his pony's slow gait, with all the disheartening appearance of a languishing invalid.

Disparate Types:
The Jagunço and the Gaucho

The southern gaucho, upon meeting the vaqueiro at this moment, would look him over commiseratingly. The northern cowboy is his very antithesis. In the matter of bearing, gesture, mode of speech, character, and habits there is no comparing the two. The former, denizen of the boundless plains, who spends his days in galloping over the pampas, and who finds his environment friendly and fascinating, has, assuredly, a more chivalrous and attractive mien. He does not know the horrors of the drought and those cruel combats with the dry-parched earth. His life is not saddened by periodic scenes of devastation and misery, the grievous sight of a calcined and absolutely impoverished soil, drained dry by the burning suns of the Equator. In his hours of peace and happiness he is not preoccupied with a future which is always a threatening one, rendering his happiness short lived and fleeting. He awakes to life amid a glowing, animating wealth of Nature; and he goes through life adventurous, jovial, eloquent of speech, valiant, and swaggering; he looks upon labor as a diversion which affords him the sport of stampedes; lord of the distances is he, as he rides the broad level-lying pasture lands, while at his shoulders, like a gaily fluttering pennant, is the inevitable scarf, or pala.

The clothes that he wears are holiday garb compared to the vaqueiro's rustic garments. His wide breeches are cut to facilitate his movements astride his hard-galloping or wildly rearing bronco and are not torn by the ripping thorns of the caatingas. Nor is his jaunty poncho ever lost by being caught on the boughs of the crooked trees. Tearing like an unleashed whirlwind across the trails, clad in large russet-colored boots with glittering silver spurs, a bright-red silk scarf at his neck, on his head his broad sombrero with its flapping brim, a gleaming pistol and dagger in the girdle about his waist — so accoutered, he is a conquering hero, merry and bold. His horse, inseparable companion of his romantic life, is a near-luxurious object, with its complicated and spectacular trappings. A ragged gaucho on a well-appareled pinto is a fitting sight, in perfectly good form, and, without feeling the least out of place, may ride through the town in festive mood.

The Vaqueiro

The vaqueiro, on the other hand, grew up under conditions the opposite of these, amid a seldom varying alternation of good times and bad, of abundance and want; and over his head hung the year-round threat of the sun, bringing with it in the course of the seasons repeated periods of devastation and misfortune. It was amid such a succession of catastrophes that his youth was spent. He grew to manhood almost without ever having been a child; what should have been the merry hours of childhood were embittered by the specter of the backland droughts, and soon enough he had to face the tormented existence that awaited him. He was one damned to life. He understood well enough that he was engaged in a conflict that knew no truce, one that imperiously demanded of him the utilization of every last drop of his energies. And so he became strong, expert, resigned, and practical. He was fitting himself for the struggle.

His appearance at first sight makes one think, vaguely, of some ancient warrior weary of the fray. His clothes are a suit of armor. Clad in his tanned leather doublet, made of goatskin or cowhide, in a leather vest, and in skintight leggings of the same material that come up to his crotch and which are fitted with knee pads, and with his hands and feet protected by calfskin gloves and shinguards, he presents the crude aspect of some medieval knight who has strayed into modern times.

This armor of his, however, reddish-gray in hue, as if it were made of flexible bronze, does not give off any scintillations; it does not gleam when the sun's rays strike it. It is dead and dusty-looking, as befits a warrior who brings back no victories from the fight.

His homemade saddle is an imitation of the one used in the Rio Grande region but is shorter and hollowed out, without the luxurious trappings of the other. Its accessories consist of a weatherproof goatskin blanket covering the animal's haunches, of a breast covering, or pectorals, and of pads attached to the mount's knees. This equipment of man and beast is adapted to the environment. Without it, they would not be able to gallop through the caatingas and over the beds of jagged rock in safety.

Nothing, to tell the truth, is more monotonous and ugly than this highly original garb of one color only, the russet-gray of tanned leather, without the slightest variation, without so much as a strip or band of any other hue. Only at rare intervals, when a "shindig" is held to the strains of the guitar, and the backwoodsman relaxes from his long hours of toil, does he add a touch of novelty to his appearance in the form of a striking vest made of jungle cat or puma skin with the spots turned out — or else he may stick a bright red bromelia in his leather cap. This, however, is no more than a passing incident and occurs but rarely.

Once the hours of merrymaking are over, the sertanejo loses his bold and frolicsome air. Not long before he had been letting himself go in the dance, the sapateado, as the sharp clack of sandals on the ground mingled with the jingling of spurs and the tinkling of tambourine bells, to the vibrant rhythms, the "rip-snortings" of the guitars; but now once more he falls back into his old habitual posture, loutish, awkward, gawky, exhibiting at the same time a strange lack of nervous energy and an extraordinary degree of fatigue.

Now, nothing is more easily to be explained than this permanent state of contrast between extreme manifestations of strength and agility and prolonged intervals of apathy. A perfect reflection of the physical forces at work about him, the man of the northern backlands has served an arduous apprenticeship in the school of adversity, and he has quickly learned to face his troubles squarely and to react to them promptly. He goes through life ambushed on all sides by sudden, incomprehensible surprises on the part of Nature, and he never knows a moment's respite. He is a combatant who all the year round is weakened and exhausted, and all the year round is strong and daring, preparing himself always for an encounter in which he will not be the victor, but in which he will not let himself be vanquished; passing from a maximum of repose to a maximum of movement, from his comfortable, slothful hammock to the hard saddle, to dart, like a streak of lightning, along the narrow trails in search of his herds. His contradictory appearance, accordingly, is a reflection of Nature herself in this region — passive before the play of the elements and passing without perceptible transition from one season to another, from a major exuberance to the penury of the parched desert, beneath the refracted glow of blazing suns. He is as inconstant as Nature. And it is natural that he should be. To live is to adapt one's self. And she has fashioned him in her own likeness: barbarous, impetuous, abrupt.

The Gaucho

The gaucho, valiant "cowpuncher" that he is, is surely without an equal when it comes to any warlike undertaking. To the shrill and vibrant sound of trumpets, he will gallop across the pampas, the butt of his lance firmly couched in his stirrup; like a madman, he will plunge into the thick of the fight; with a shout of triumph, he is swallowed up from sight in the swirl of combat, where nothing is to be seen but the flashing of sword on sword; transforming his horse into a projectile, he will rout squadrons and trample his adversaries, or — he will fall in the struggle which he entered with so supreme a disregard for his life.

The Jagunço

The jagunço is less theatrically heroic; he is more tenacious; he holds out better; he is stronger and more dangerous, made of sterner stuff. He rarely assumes this romantic and vainglorious pose. Rather, he seeks out his adversary with the firm purpose of destroying him by whatever means he may. He is accustomed to prolonged and obscure conflicts, without any expansive display of enthusiasm. His life is one long arduously achieved conquest in the course of his daily task. He does not indulge in the slightest muscular contraction, the slightest expenditure of nervous energy, without being certain of the result. He coldly calculates his enemy. At dagger-play he does no feinting. In aiming the long rifle, he "sleeps upon the sights."

Should his missile fail to reach its mark and his enemy not fall, the gaucho, beaten or done in, is a very weak individual in the grip of a situation in which he is placed at a disadvantage or of the outcome of which he is uncertain. This is not the case with the jagunço. He bides his time. He is a demon when it comes to leading his enemy on; and the latter has before him, from this hour forth, sighting him down his musket barrel, a man who hates him with an inextinguishable hatred and who lies hidden there in the shade of the thicket.

The Vaqueiros

These contrasting characteristics are prominent in normal times.

Thus, every sertanejo is a vaqueiro. Aside from the rudimentary agriculture of the bottom plantations on the edge of the rivers, for the growing of those cereals which are a prime necessity, cattle-breeding is in these parts the kind of labor that is least unsuited to the inhabitants and to the soil. On these backland ranches, however, one does not meet with the festive bustle of the southern estancias.

"Making the roundup" is for the gaucho a daily festival, of which the showy cavalcades on special occasions are little more than an elaboration. Within the narrow confines of the mango groves or out on the open plain, cowpunchers, foremen, and peons may be seen rounding up the herd, through the brooks and gullies, pursuing intractable steers, lassoing the wild pony, or felling the rearing bull with the boleador, as if they were playing a game of rings; their movements are executed with incredible swiftness, and they all gallop after one another, yelling lustily at the top of their lungs and creating a great tumult, as if having the best time in the world. In the course of their less strenuous labors, on the other hand, when they come to brand the cattle, treating their wounds, leading away those destined for the slaughter, separating the tame steers, and picking out the broncos condemned to the horsebreaker's spurs — at times like these, the same fire that heats the branding iron provides the embers with which to prepare the roast, cooked with the skin on, for their rude feasts, or serves to boil the water for their strong and bitter-tasting Paraguayan tea, or mate. And so their days go by, well filled and varied.

Unconscious Servitude

The same thing does not happen in the north. Unlike the estancieiro, the fazendeiro of the backlands lives on the seaboard, at a distance from his extensive properties, which he sometimes never sees. He is heir to an old historic vice. Like the landed proprietor of colonial days, he parasitically enjoys the revenues from vast domains without fixed boundaries, and the cowboys are his submissive servants. As the result of a contract in accordance with which they receive a certain percentage of what is produced, the latter remain attached to the same plot of ground; they are born, they live and die, these beings whom no one ever hears of, lost to sight in the backland trails and their poverty-stricken huts; and they spend their entire lives in faithfully caring for herds that do not belong to them. Their real employer, an absentee one, well knows how loyal they are and does not oversee them; at best, he barely knows their names.

Clad, then, in their characteristic leathern garb, the sertanejos throw up their cottages, built of thatch and wooden stakes, on the very edge of the water pits, as rapidly as if they were pitching tents, and enter with resignation upon a servitude that holds out no attractions for them. The first thing that they do is to learn the ABC's, all that there is to be known, of an art in which they end by becoming past masters — that of being able to distinguish the "irons" of their own and neighboring ranches. This is the term applied to all the various signs, markings, letters, capriciously wrought initials, and the like which are branded with fire on the animal's haunches, and which are supplemented by small notches cut in its ears. By this branding the ownership of the steer is established. He may break through boundaries and roam at will, but he bears an indelible imprint which will restore him to the herd to which he belongs. For the cowboy is not content with knowing by heart the brands of his own ranch; he learns those of other ranches as well; and sometime, by an extraordinary feat of memory, he comes to know, one by one, not only the animals that are in his charge but those of his neighbors also, along with their genealogy, characteristic habits, their names, ages, etc. Accordingly, should a strange animal, but one whose brand he knows, show up in his herd, he will restore it promptly. Otherwise, he will keep the intruder and care for it as he does for the others, but he will not take it to the annual fair, nor will he use it for any labor, for it does not belong to him; he will let it die of old age.

When a cow gives birth to a calf, he brands the latter with the same private mark, displaying a perfection of artistry in doing so; and he will repeat the process with all its descendants. One out of every four calves he sets aside as his own; that is his pay. He has the same understanding with his boss, whom he does not know, that he has with his neighbor, and, without judges or witnesses, he adheres strictly to this unusual contract, which no one has worded or drawn up.

It often happens that, after long years, he will succeed in deciphering the brand on a strayed bullock, and the fortunate owner will then receive, in place of the single animal that had wandered from his herd, and which he has long since forgotten, all the progeny for which it has been responsible. This seems fantastic, but it is nonetheless a well-known fact in the backlands. We mention it as a fascinating illustration of the probity of these backwoodsmen. The great landed proprietors, the owners of the herds, know it well. They all have the same partnership agreement with the vaqueiro, summed up in the single clause which gives him, in exchange for the care that he bestows upon the herd, one-fourth of the products of the ranch; and they are assured that he will never filch on the percentage.

The settlement of accounts is made at the end of winter and ordinarily takes place without the presence of the party who is chiefly interested. That is a formality that may be dispensed with. The vaqueiro will scrupulously separate the large majority of the new cattle (on which he puts the brand of the ranch) as belonging to the boss, while keeping for himself only the one out of every four that falls to him by lot. These he will brand with his own private mark, and will either keep them or sell them. He writes to the boss, giving him a minute account of everything that has happened on the place, going into the most trivial details; and then he will get on with his never interrupted task.

That task, although on occasion it can be tiring enough, is an extremely rudimentary one. There does not exist in the north a cattle-raising industry. The herds live and multiply in haphazard fashion. Branded in June, the new steers proceed to lose themselves in the caatingas along with the rest. Here their ranks are thinned by intense epizootic infections, chief among them rengue, a form of lameness, and the disease known as the mal triste. The cowboys are able to do little to halt the ravages of these affections, their activities being confined to riding the long, endless trails. Should the herd develop an epidemic of worms, they know a better specific than mercury: prayer. The cowboy does not need to see the suffering animal. It is enough for him to turn his face in its direction and say a prayer, tracing on the ground as he does so a maze of cabalistic lines. And, what is more amazing still, he will cure it by some such means as this.

Thus their days are spent, full of movement, but with little to show for it all. Rarely does any incident, some slight variation, come to break the monotony of their life.

Bound together by a spirit of solidarity, they unconditionally aid one another at every turn. Let a giddy steer flee the herd, and the vaqueiro will snatch up his wooden lance put spurs to his nag, and gallop after it in hot haste. If his efforts do not meet with success, he has recourse to his neighboring companions and asks them to "take the field," a phrase that is characteristic of these rustic knights; and his hard-riding, lusty-yelling friends by the dozen and the score will then follow him, scouring the countryside, riding over the slopes and searching the caatingas, until the beast, in the language of the cattlemen, has been "taken down a peg" and "turns up his nose" or else is thrown by main force when his horns are grasped in the cowboys' powerful hands.

Glossary of Regional Terms

Bandeira   An armed band in colonial days, composed of adventurers, particularly those of the São Paulo region, who made their way into the backland.
Bandeirante   A member of a *bandeira*.
Caboclo   Brazilian Indian (feminine: *cabocla*).
Cafuso   Offspring of Indian and Negro parents.
Campos Gerais   Open country or country that is not forested; applied in particular to the highland plains of Paraná.
Corredeira   A stretch of river marked by a series of small waterfalls.
Curiboca   Offspring of white and Indian parents (cf. *mameluco*).
Estancia   A country estate or ranch in southern Brazil.

Fazenda   A ranch, plantation, or ranch house, especially of northern Brazil.

Gaucho   A cowboy of southern Brazil (cf. *vaqueiro*).

Jagunço   A ruffian; used as synonymous with *sertanejo*.

Mameluco   Offspring of white and Indian parents (cf. *curiboca*).

Mestizo   Any racial hybrid (broad sense): otherwise, a White-Indian hybrid.

Sertanejo   Native of the *sertões* or backlands.

Sertão (pl. *sertões*)   The interior of the country, backlands; the term applied particularly to the region in northeastern Brazil centering in the state of Baía.

Vaqueiro   A cowboy of northern Brazil (cf. *gaucho*).

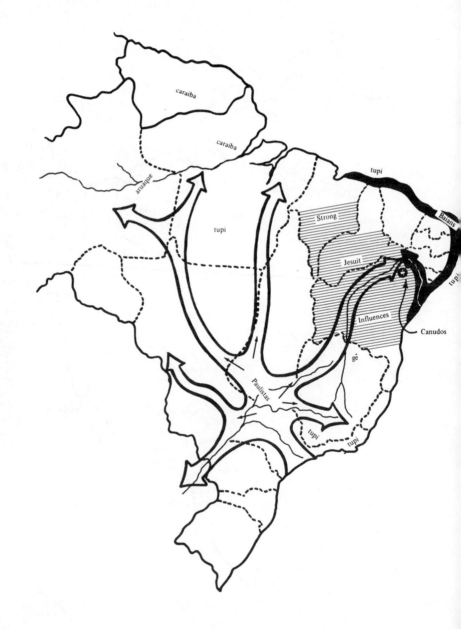

On the Biology of Race:
An Essay in Three Parts

1. Genetic variation
within local populations

The local population of this essay is the collection of individuals who live within a spatially restricted area and whose mates are chosen for the most part from within that area. For man the local population is the village, the parish, or in large cities, the neighborhood. Even in regions where a species appears to be distributed continuously over enormous areas, individuals are gathered together here and there in small batches and these are the local populations. In the case of human populations, a large fraction of all marriages are between couples whose birthplaces are very close to one another – one or two miles. Even the automobile has failed to extend this distance by much. Most fruit flies mate with others born at distances less than several yards away.

The individuals of a local population differ genetically from one another. We recognize in the case of man that no two persons are identical. Ten or twenty years ago the dissimilarities between persons were regarded as superficial – genetic in origin, to be sure, but representing only a small fraction (one per cent or less) of the entire set of human genes. Today our view has changed considerably. It is quite possible that of all pairs of genes found in an "average" person, as many as two of every five may consist of dissimilar members; within the human community approximately one-half of all gene loci seem to be represented by two or more common mutant genes.

The existence of individual-to-individual variation is not unique to mankind. Equally large differences are found between individual flies of many species. It is only our lack of familiarity with species other than man that makes it difficult to see dissimilarities; it takes a cat-lover or a dog-lover to appreciate individual differences in these animals. Indeed, every species of animal that has been examined by modern analytical techniques has been found to possess a wealth of genetic variation not previously suspected. Uniformity between the individual members of a species is an illusion, at least at the molecular level.

General agreement on the means by which genetic variation is maintained in local populations has not yet been reached. Its existence, however, is no longer in dispute. Nor is its origin, because mutation is a well-known genetic phenomenon; faulty replication of genetic material (DNA) will cause a gene to produce mutant forms of itself at rare but more-or-less regular intervals. The disagreement concerning existing variation is whether recurrent mutation is frequent enough to account for the amount observed in populations or whether certain combinations of dissimilar genes are more likely to reproduce than others and thereby maintain the genetic variation in the population. Recurrent mutation, the sporadic alteration of a normal gene to a mutant form, tends to pump mutations into a large population of cross-fertilizing individuals; the mutations are pumped out of the population again through the lowered viability of mutant individuals, or their lack of reproductive success. The combined effect of these two processes – one, putting mutations into the population; the other, removing them – is to establish a rather constant frequency of mutations within the population. On the other hand, the observed frequency of "mutant" genes could be accounted for equally well if individuals possessing two dissimilar alleles were to enjoy greater reproductive success than those individuals carrying two identical ones.

That genetic variation exists in local populations is important to us for this essay; *why* it exists is not. The point we want to make, and to make as strongly as possible, is that there are no two persons (with the exception of identical twins) who are genetically identical – neither neighbors in a community, nor father and son, nor brothers, nor sisters. We sometimes read about this or that human "stock" as if it were a collection of identical individuals. This simply is not so. No two individuals of any crossbreeding species are alike, let alone an entire "stock" of individuals. Only potatoes and other plants that are maintained by cuttings exist as "stocks." The number of gene loci at which two different forms of gene are found is so large, it is unlikely that any two of the billions of sperm produced by one man are identical. That is the extent of genetic variation in man, and in this respect, man is very much like all other animal species.

2. Genetic variation
between local populations

No two individuals of a population are identical. This claim is based on the tremendous number of genes for which alternative forms (alleles) exist and the exceedingly small chance that any two sets of genes chosen from those available to an interbreeding population would be identical in all respects. Even in bridge, where there are only thirteen cards of each of four suits, the chance that a player will receive the same hand in two successive games is extremely small (approximately one chance in 650 billion). Even the chance that a player will receive precisely the same hand twice in a lifetime of bridge playing is exceedingly small. The "deck" of genes from which individuals draw their hands − one hand from the mother, one from the father − is a complicated one; there are thousands of genes rather than thirteen and the number of mutant forms (= suits) that each can take is much greater than four.

What, then, are the differences between populations that inhabit different localities? How can individuals that differ already from one another differ even more from individuals of another population? The answers lie both in the actual genes that differ and in the chance that certain combinations of genes will arise in each population. The individuals of successive generations of a particular local population represent hands that are dealt from the same deck, a deck that is shuffled each generation prior to the formation of germ cells. These individuals may differ but they differ in understandable ways because they are all chosen from the same pool of genes; vary as they may from deal to deal, bridge hands contain few genuine surprises. Individuals of different populations are put together from different pools of genes; it is as if they were dealt from different decks that need not closely resemble one another.

The gene pool of a population is the total assemblage of genes carried by the individual members of that population. One can imagine collecting all of the sperm of adult males in one container and all of the mature eggs of adult females in another. For a population of as few as several hundred individuals, this would mean billions of sperm and tens of thousands of eggs; the genes of these two containers would represent (literally) the pooled genes of the population (its "gene pool"), and it is from chance combinations of one sperm drawn from the one container and one egg drawn from the other that new individuals are made.

The gene pools of geographically isolated populations, even though the original pioneers that started each may have been drawn from a single locality, will in time come to differ. The differences arise in part by chance, but for the greater part they arise because of the different environments in which the two populations live and the different courses the populations take in response to local selection pressures. Differences between flies collected at different geographic localities have been shown repeatedly to have their bases in genetic differences. The proportion of eggs that hatch and develop successfully into adult flies depends upon the temperature at which the experimental cultures are kept. Different species of fruit flies have different optimal temperatures. Furthermore, offspring of flies of the same species caught at different geographic localities also have different optimal temperatures; those caught in Moscow, for example, develop better in low temperatures than do corresponding flies captured in Egypt.

The origin of genetic differences as the result of "natural" selection has been observed in laboratory populations. A striking change may take several years to arise, although this statement depends upon what we call "striking." As a rule, flies captured in the cooler parts of a country (northern as opposed to southern in Europe and North America; high as opposed to low elevation) are larger than those captured in the warmer regions. This is true not only of the flies caught in the field but also of their offspring raised in the laboratory. Flies of cool and warm regions *differ genetically* in respect to size. In one series of laboratory populations kept at different temperatures for an extended time, no genetic difference of this sort was discernible after eighteen months whereas a pronounced difference was detected after six years.

The mechanism by which selection works to alter the composition of the gene pool of a local population is through the unequal reproductive rates of individuals that differ genetically. The reproductive advantage of the carriers of different mutations may lie in their greater survival during preadult life or a greater fertility after reaching maturity or both. In transmitting his genes to future generations, a sterile adult is no more effective than is a dead juvenile.

The points made so far in this essay may be summarized as follows. Individual genes may take on a variety of forms through mutation. Within any one locality, mutant genes exist in high frequency at a large proportion of all gene loci and, hence, enormous numbers of gene combinations are possible. The number of possible gene combinations is so large that no two individuals (except identical twins) in the history of the earth have ever been genetically identical.

Different populations, under the influence of local selective forces, come to possess different arrays of mutant genes at different sets of gene loci. Consequently, not only do individuals of the same population differ from each other but also individuals from one population differ systematically from those of another.

What we have said for local populations so far applies to any two isolated populations. Two populations of flies living along different stream beds in a desert area, mice living in different meadows, or, indeed, persons living in two communities such as Towanda, Pennsylvania, and Youngstown, Ohio, come to differ from one another. The following essay will deal with populations that have been isolated for considerably greater lengths of time.

3. Races and racial differences

Individuals differ from each other and, as we have just seen, local populations differ from one another as well. The differences between populations tend to grow larger the longer they are separated from each other. The Dunkards, a religious sect originally from Germany and now living (in part) in Pennsylvania, were presumably a representative sample of the German population in which the sect arose. By restrictive marriage customs, the sect has isolated itself from its neighbors and, in respect to the frequencies of genes controlling blood groups, shapes of ear lobes, and other physical traits, now resembles neither the Germans of the area from which its members originally came nor the Pennsylvanians among whom they now live.

At some point in time differences between populations become great enough that persons, simply as a matter of convenience, lump under a single term those that exhibit a certain characteristic in common. Thus, races are born. The point about *convenience* must be repeated once more. A racial designation is a matter of convenience and nothing more. What is convenient for one person may not be convenient for others and so a listing of races – human or otherwise – depends largely upon the author of the list. The persons that live in and around Towanda, Pennsylvania, as I mentioned in the previous essay, could surely be shown to differ genetically as a group from those in Youngstown, Ohio, but no reasonable person would refer to these two populations as races. On the other hand, there are obvious and systematic differences between the people of Great Britain, Scandinavia, and Italy. There are anthropologists who have divided the people of Europe into numerous races, but there are others who merely lump Europeans with other "white" races under the label Caucasian.

For any species but man, and maybe for primitive man as well, race is both a geographic concept and a population concept. It is a geographic concept because races are constellations of local populations that have diverged from other constellations, and divergence takes place only in isolation – geographic isolation for the most part. Only man can devise means whereby populations can work side by side, go to school side by side, and play side by side but not intermarry. For any other species of plant or animal, individuals that inhabit a certain small area constitute *the* local population of that area.

Race is a population concept; this is a point that must be repeated over and over because it is a point that must be understood. No two individuals of a population are identical nor are any two members of a race identical. There is no truth to the old notion of a "pure" race; pure races do not exist and anyone who thinks they do is thinking nonsense.

Because all members of a population are genetically different, the population must be thought of as embracing all of the various types of individuals that are generated within it by the chance matings of its members. Since race is a population concept, a race must embrace all of the various kinds of individuals that are put together within it by chance. If we were to talk about an Anglo-Saxon race, for example, we would be discussing for the most part fairskinned, blondish persons who (to listen to them) suffer unspeakably if exposed to sunlight or to summer temperatures greater than 65°F but we would be discussing as well those rarer dark-skinned, black-haired individuals who also appear in British families. The population – and the race – includes the whole gamut of dissimilar individuals who are generated by the local mating (or marriage) patterns.

There is enough genetic variation within any one human race so that a clever genetic surgeon could assemble genes from it and in this way create a person who would closely resemble one commonly found within any other race. Despite appearances, however, this artificially assembled person would belong to the race from which he was generated. If our surgeon could assemble the proper genetic material by picking and choosing from the material at hand, chance would eventually assemble the same combination. And when the person arises as one of the gamut of all persons produced in his particular generation, he will belong with the race that gave rise to him. So much, then, for "black" races and "white" races. There are black men and white men, that is, men whose skin colors differ, but these persons are not races. Races are groups of interbreeding individuals, populations. The members of a race are those persons who are born to it; this is so irrespective of their individual characteristics.

Not long ago a South African child was removed from his parents and family because they were officially classified as "white" whereas he was dark enough to be classified "coloured." It is tragic that persons feel the need to take classifications of this sort seriously; that is surely the enormous tragedy that engulfed this South African family as well as countless other persons during the history of the world. On the other hand, if a classification based on physical appearance is used in defining "race," families will be split over and over again wherever races are segregated. The problem will arise whether the group in power is white and blacks are excluded, or whether these in power are black and whites are excluded. If race is not recognized as a population concept, which must include the family as a unit within the population, and if physical characteristics (whether they are those evaluated on sight or those requiring the careful measurements of a trained anatomist) are used instead, families will be split. Mendel made this point clear more than 100 years ago.

The Role of Labels in Describing and Understanding Evolving Systems

"And remember how they were all the time taking snapshots and asking for glasses of water? And their funny clothes? And how loud they were? But generous. Generous to a fault." The speaker is a proprietor of a small French café; he and his wife are shown in a *New Yorker* cartoon by Opie. They are reminiscing over the vanished American tourist while standing before their empty restaurant, staring at an empty street. The ice water, the funny clothes, the camera, and the generosity — Opie has emphasized a cluster of traits that evoke an instantaneous vision of the missing tourist. Pushed to continue his thoughts on Americans, the café owner might not hesitate to use the word "race" in identifying the people of whom American tourists are a part. If he were to use the word "race" in continuing his reminiscences, however, he would be using the word very differently than does the geneticist. Its use in reference to Americans would be confined to characteristics instilled in a people by an American culture. Biologically, Americans are a diverse people. Our origins lie in all parts of the globe and, imperfect as the melting pot has been, a bewildering variety of otherwise unlikely marriages have occurred within the United States. The American tourist — cold-water lover, camera fan, and the rest — may carry genes whose progenitors arose on several different continents. To whatever extent Americans share common traits by which their nationality can be recognized by strangers, these traits are cultural in origin.

Culture plays a more important part in the recognition of other peoples than many students of mankind could care to admit. In many scholarly books dealing with the races of man the reader is confronted with photographs or drawings of persons from faraway lands. These illustrations are intended to show representatives of many different races. We note in these pictures the physical characteristics such as head shape and facial features, but equally impressive are the hair styles, necklaces, headdresses, shirts (or lack of them), and other clothing. The tribal hairdos adorned with mud and dung are as characteristic as any common physical feature exhibited by a population of African natives.

The accounts of vaqueiros and gauchos given in *Rebellion in the Backlands* reveal the impact contrasting environments and cultures have on isolated societies. When Cunha describes the gait of the vaqueiro as "unsteady, slightly swaying, sinuous" or the man himself as "ugly, awkward, stooped" he is not describing biological aspects of the vaqueiro; he is describing the mien of a man eking out an existence under the most marginal conditions of life, a man who spends long hours on horseback, and a man who by tradition takes full advantage of rare moments of rest by an utter physical relaxation – much as a seasoned soldier relaxes during his ten-minute break when on patrol.

Cultural traits imposed on men by the society in which they are raised or by the environment in which they live are not always easily separated from traits that have a truly genetic basis. By their nature, both tend to run in families. At one time the physical and mental retardation of Southern children that was caused by hookworm infections was interpreted as a genetic disorder. *Kuru,* a lethal neurological disorder caused by a strange virus-like particle and seemingly transmitted by cannibalism, was until recently also interpreted as a genetic disease. Data on the occurrence of rheumatic fever, a complication known to follow steptococcus infections, can be interpreted in genetic terms. Under circumstances such as these, it is easy to forgive the innocence of Euclides da Cunha for interpreting cultural features as biological ones.

The identification of cultural characteristics as racial ones is only a small step beyond noting the correlation between cultural traits and certain physical traits of a people. Then the next small step follows, that of identifying groups of persons by means of particular physical characteristics. Consequently, the photographs and drawings of representative individuals that are often found in textbooks on human evolution must be treated gingerly. First, as I pointed out above, each picture generally includes a host of unobtrusive cultural features that serve to identify the subject's origin. More important, perhaps, is the need to remember the order of importance in the sequence: population, individual, and photograph. Our thoughts about these should go in the following sequence. (1) Here is a population that occupies a certain geographical region and seems to be reasonably self-contained in respect to marriage ties. (2) There are certain physical features that are especially common in this population and, indeed, there is a gentleman who might serve as a representative model for illustrative purposes. (3) Here is the gentleman's photograph in the textbook: Figure 28. An example of the Sinopolynesian race of Outer Tumi Island.

The existence of a photograph in one's hand and the remoteness of the population itself tempt us to establish an alternative (and incorrect) train of thought. (1) Here I see a typical (notice that I did not say "representative" this time) Sinopolynesian. I can see that he has closely cropped hair that is parted on the left, a moustache, and a nose that is small and slightly awry. Furthermore, he appears to be rather dark-skinned (printer's mistake?) and slight in build. (2) All Sinopolynesians must have closely cropped hair that is parted on the left, moustaches, and so forth. (3) Any person who has closely cropped hair that is parted on the left, a moustache, and so forth must be Sinopolynesian.

I have gone through this account of the fictitious Sinopolynesian in a deliberately ludicrous way in order to illustrate, by emphasis, the flaw in the race concept held by early physical anthropologists. These persons made measurements on persons in all continents, they set up racial classifications based on their measurements, and they then assigned persons who fit a given measurement into that particular race. Thus, a pocket of individuals in central Ceylon might by chance correspond in some way to another group on the west coast of Africa; despite their geographic isolation, these groups would fall within the same racial classification.

There is nothing inherently wrong with a classification of objects — books or persons — based on physical measurements; the flaw lies in what is done with the measurements and in the interpretation of the classification to which they lead. The flaw, in short, is in the eye of the beholder.

The classification of books in a library illustrates much of what I want to say about the classification of people. Books can be classified and stored according to many schemes: by their size, by the color of their covers, by subject matter, or alphabetically by title or author. Most librarians use the Dewey system, one that groups books according to subject matter. The intense browser profits by this scheme because he can search for a book on a particular topic without actually knowing that a suitable book really exists. Bookstores, unlike libraries, often arrange their books by authors because the customer who comes to buy usually knows what he is after. Classification by size is a space-saving device; libraries will be forced to this system very quickly if the present rate of publication continues. Color offers no advantages as a means of classification; it saves no space because books of the same color may be of quite different sizes, it offers no hint as to content or author, and it itself is liable to change through fading or re-covering.

The reason for classifying books is to organize their storage in a manner that permits the rapid retrieval of individual books upon demand. Any orderly scheme will work. The system becomes nonsensical only when persons use the classification system for purposes for which it was not intended. A college course that requires the student to buy four books by Hemingway or two books on conservation is not at all strange. One that requires him to buy three 12-inch books or a half-dozen blue books would be bizarre, to say the least. This is precisely the type of misuse that is frequently made of anthropological measurements.

Taxonomic classifications are fundamentally schemes for storing information regarding man's knowledge of the existence of other organisms; nevertheless, they seem invariably to evolve into schemes reflecting evolutionary relationships. Related species and related populations of men do tend to resemble one another. This fact does not mean, however, that species or populations resembling one another in a specified way are necessarily closely related. During the 1930's one large publishing house brought out many scientific textbooks, each of which had a depressingly dull brown cover. We could predict with near certainly that the next scientific textbook in that series would also have a brown cover, and we could even predict with some assurance (because the series was a large one) that a particular brown-covered book would be a member of this series. There were, however, enough drab, brown-covered books that were not members of the series, that were not even science books, to make nonsensical the notion that all brown textbooks were science books.

The notion that a race of people defined according to one or more physical measurements is a group of persons who are in some realistic sense related to one another through recent common ancestry is as nonsensical as the notion that brown-covered books are science books. With the growing emphasis on tracing evolutionary lines of descent − on disentangling, as Euclides da Cunha has attempted to do, the migration routes and patterns of interbreeding of groups and populations of persons − the absurdity of stretching *taxonomic* racial designations in an attempt to encompass truly *biological* racial concepts has become increasingly clear. The result has been that many modern physical anthropologists deny that races exist. In reality, they mean that those categories called races by their predecessors are not useful categories under today's research patterns and research interests. They are not denying, to use an exaggerated example, the existence of persons whose heights fall between five-foot-five and five-foot-seven; they are merely saying that it serves no useful purpose to know who these persons are, and it serves an even less useful purpose to refer to these persons as a race − as the five-foot-six race.

Labeling individuals according to brown, black, white, red, and yellow races is no more useful for today's concepts of man, his origins, and his descent through time than is placing them is a series such as under-five-feet, five-foot-two, five-foot-four, five-foot-six, five-foot-eight, and so forth. Although related persons and related populations might *tend* to fall within a single classification, the numerous exceptions (both relatives that fall outside and strangers that fall inside) make impossible the use of this classification for purposes of depicting relationships between peoples. The absurdities come from erecting an image and then trying to make a segment of mankind fit it instead of selecting a segment of mankind and then attempting to devise a means for referring to it while talking and writing.

The ordinary person today − "the man in the street" − has a concept of race that is based on a mixture of cultural characteristics and physical attributes. It is a strange, worthless, and at times vicious concept. Any physical trait that is associated with a given population becomes a label so that wherever it is found, the person carrying it must, by (irrational) definition, be of *that* population. A characteristic fold of the eyelid identifies an "Oriental" whether the person lives in mainland China, is someone who had a remote ancestor of Chinese descent, or merely carries the gene for eye-lid folding − a gene found in many human populations.

A small controversy has developed around the word "race." Those persons whose training has emphasized the futility in attempting to learn of man's evolution (and of the interrelations between peoples) by forcing it to conform with arbitrary labels denounce racial designations. They would substitute instead "ethnic group" or some other new term that would embrace cultural and linguistic differences between persons as well as occasional physical differences. Those whose training has led them directly to look upon populations as evolving entities and who have sought a term by which they could refer to clusters of related populations have used and continue to use the word "race." A new word could, of course, be invented, but it would continue to have the meaning now assigned to "race" under present concepts. In the case of persons who persist in attaching greatest importance to the label and who insist that people be made to fit labels rather than the reverse, the substitution of "ethnic group" for "race" will help very little. Cultures and languages, like populations, evolve. This evolution will not be revealed by lumping similar cultures and languages under one label; nor will the use of a label halt cultural and linguistic evolution. Labels, whether racial or ethnic, are merely aids to communication and must serve the objects that are under study; evolving people, evolving cultures, and evolving languages are the "realities" under discussion.

A Potpourri
of Racial Nonsense

Seldom does man have such an opportunity to indulge in vindictive nonsense as he does in matters of race. Furthermore, only racial affairs are so muddled that nonsense emanates with equal frequency from both sides of a controversy. "As a result of doctrines embodying the concept of race, but having no scientific foundation," writes William C. Boyd, "millions of people have been tortured or killed. Yet nearly everyone, even the *victims* of this delusion, seemed to assume that distinct, physical races do exist and that each possesses well-defined differences in mental abilities and attitudes."* We ordinarily expect the "Logic of Rightousness" to combat and triumph over the "Foolishness of the Wicked"; in racial discussions, nonsense seems to be pervasive and endless. The truly encouraging signs lie in the view blacks have adopted of themselves. Controversies about race are, for the most part, controversies among whites; the bright and articulate blacks have clearly seen this and emphasize it pointedly in their public statements. The discouraging signs, on the other hand, are those suggesting that many blacks themselves have finally succumbed to the fallacies that have plagued whites for so long. The purpose of this essay is to list, with appropriate comments, the absurdities most frequently encountered in discussions of racial matters.

There Is No Such Thing as Race. . . .

Race, as I have used the term in the two preceding essays, is a population or a collection of geographically contiguous populations that differ genetically in a systematic manner from other populations or collections of populations. Because all local populations differ from one another to some extent, the racial designation is given only as a matter of convenience; different persons find different numbers of races and different patterns of subdivision convenient for their purposes. Dissimilar classifications by various authors lead to confusion, but this does not mean that the distinctions upon which a given classification is

---

*William C. Boyd, *Genetics and the Races of Man* (Boston: D. C. Heath and Co., 1950), p. 185.

made do not exist. If the word "race" were not used in these instances when groupings are convenient, another word with precisely the same meaning would be invented.

Races have no permanence; it is the species, not the race, that is reproductively isolated from other groups and therefore unable to draw on genetic material other than its own in meeting evolutionary demands. In contrast to the lack of crossbreeding between species, races arise and then fade away once more while new ones form according to new geographic patterns. Any attempt to assign a permanent name to a race is doomed to failure because the group of organisms being named has no permanance. This does not destroy the usefulness of applying a temporary label of convenience, and this label is a racial designation.

An Individual's Characteristics
Determine the Race to Which He Belongs. . . .

Individual characteristics cannot be used in racial classification if *race* is to refer to populations of individuals. The dark-skinned Englishmen does not belong to a Mediterranean race nor is the blonde Italian a Nordic. Populations generate a variety of individuals just as a deck of bridge cards generates a variety of hands; each of the entire gamut of individuals produced by the mating of the members of a population belong to that population and to the race to which that population is assigned.

To adopt the contrary view, that is, to classify an individual by his personal characteristics, leads to the indefensible notion that members of the same family can belong to different races. This is entirely wrong. A person belongs to a given race because his parents, his grandparents, and even more remote ancestors belonged to a particular group of interbreeding individuals. Race is a matter of pedigree, not of appearance. Geneticists have a perfectly good word that refers to the personal characteristics of individuals: *phenotype.* Race is not to be confused with phenotype.

Negro: A Black Man;
One Who Has More or Less Negro Blood. . . .

This is a dictionary definition of Negro. Insofar as Negro is one of the races of man, the definition errs on at least three points. First, it refers to "black man"; black is not a race nor is a black man a member of the Negro race because of his blackness. Skin color is merely one aspect of a person's appearance (his *phenotype*). It has nothing to do with his racial classification – a classification that must be made on the basis of a population of individuals. Second, the definition drags in "blood," a marvelous substance that is necessary for life but one that has nothing to do with either race or heredity. Blood carries oxygen from the lungs to the tissues of the body and performs other physiological tasks; it has no genetic function. Third, the concept of "Negro-ness" expressed in the definition is more nearly a religious than a biological or scientific one. I make this claim recalling that in the "mixed" marriage of a Roman Catholic and a non-Catholic, the children are regarded as Catholic. This is virtually the attitude on racial designation held in the United States at the present time; a Negro *is defined* as one who has in his pedigree at least one ancestor of African descent. Under this view, the proportion of Negroes in the United States must increase each generation by an amount that depends upon the proportion of interracial marriages. The converse of the same erroneous view convinces the race-conscious Negro that his people are being "assimilated" by whites; in his eyes, the proportion of Negroes *decreases* each generation because of black-white intermarriages. A definition of race that leads to diametrically opposed conclusions depending upon one's point of view is worthless – or worse.

Each Person Carries
Within Him His Racial Average. . . .

This particularly vicious claim is seemingly designed to deny hotel reservations, club memberships, and high-priced suburban homes to blacks who cannot be excluded on economic grounds.

The racial average generally cited by those who cite such things is IQ score. Numerous tests given to Negroes and whites in schools throughout the country for a half-century or more have shown that the average score attained by Negroes is some ten or fifteen points below that of whites. The meaning of this difference is the subject of my next essay. Right now the question concerns the assignment of this lower *average* to each individual Negro despite his *personal* IQ rating. If the same practice were followed in respect to weight, no person would be overweight or underweight because each individual would be assigned his racial average. If blacks on the average weigh somewhat less than whites, then no individual black would be acknowledged to weigh more than a white person.

The main purpose of the "racial average" claim is to perpetuate the notion of racial superiority. If the trait (such as intelligence) in question can be sold as one for which differences reflect superiority and inferiority, and if Negroes can be shown to have the inferior average, a conviction that he carries the hidden average of his superior race can be a powerful morale builder for a low-scoring white. Conversely, if he can be persuaded of the validity of that argument, it can destroy the morale of even the most intelligent black.

The notion that a racial average is a thing that one carries — as a halo or as an albatross, depending upon one's group — is preposterous. Each individual is the product of the complex interactions of his environment — cultural, social, and physical — and his genotype. There is no mysterious average lurking within each of us to be passed on to our children; there is no average upon which to call in order to inflate our own worth or denigrate that of others.

The Intelligence of Negroes
Is Inferior to That of Whites

The next essay deals with this claim from the point of view of testing procedures and the interpretation of experimental observations. At the moment, I merely call attention to what appear to me to be discrepancies in the evidence generally cited to support the claim that Negroes are genetically inferior to whites in intellect. Analyses of several different genes affecting blood groups and other inconspicuous genetic traits have shown that about one-third of the hereditary material of American Negroes has come from the numerically larger white population. Negro children, according to some educational psychologists, consistently score lower on IQ tests than do whites. Studies of genetically segregating populations of Negro-white racial hybrids, according to Dr. A. M. Shuey, are not particularly useful in the analysis of racial intelligence.

If race (that is, ultimate geographic origin) made any substantial genetic contribution to intelligence, American Negroes should, in my opinion, yield highly variable rather than consistent results. Moreover, this variation should be especially conspicuous in genetically segregating populations. That the study of precisely these segregating populations is dismissed as unrewarding appears on the face of it to argue against any important relationship between intelligence and racial (or, more accurately, geographic) origin of genetic material.

An important feature of the genetics of racial intelligence has been summarized by Dr. Joshua Lederberg in these words: " 'Intelligence' undoubtedly does have a very large and relatively simple genetic component. In fact, the genes are all too visible, they control the color of the skin." This statement offers an explanation for the discrepancies noted above. Segregating genes do not cause IQ scores to vary because the only genes that have a marked effect are those that are externally visible and, hence, govern the interactions of their bearers and the society within which they live. There is no compelling evidence, as we shall see in the following essay, to cause us to believe that in determining intelligence, genes without these external and highly prejudicial societal effects have any systematic effect on IQ that is related to their racial (that is, geographic) origins.

Race and IQ:
A Critique of Existing Data

The exposure of a bacterial culture to ultraviolet light leads to a marked increase in the number of mutant individuals; it also results in the death of some of the exposed individuals. There was a time not too long ago when neither of these events was well understood. The exposure of bacterial culture to ultraviolet light and then to daylight or artificial "white" light reduces the proportion of "dead" bacteria; this phenomenon is known as photoreactivation (the reactivation of "dead" bacteria by light). There was a time when this particular phenomenon was not understood either. Following a seminar in which a visiting lecturer had admitted, first, that his data on ultraviolet exposure had failed to explain the mechanism by which mutations are induced in bacteria and, second, that additional experiments involving photoreactivation had also failed to clarify matters, I visited the home of a colleague who had also attended the lecture. "How," he asked sarcastically, "can that man expect to resolve one unknown by the use of a second one?"

In many respects the educational psychologists who study the relative IQ's of different races find themselves repeating the mistake of the visiting lecturer. Race is such a confused and maligned concept that a fair number of scientists claim that there is no such thing; races, to these persons, simply do not exist. On the other hand, Intelligence Quotient (IQ) is such a poorly defined quantity that there is no general agreement, even among psychologists who are responsible for its measurement and continued use, as to what it does or does not measure. The study of the relative intelligence of different races, like the lecturer's poorly designed experiment on mutation in bacteria, appears to confound confusion in the hope that something understandable will emerge.

Many studies purporting to demonstrate the lower IQ of Negroes relative to that of whites have been brought to public attention in recent years. Such studies have been cited vociferously and, of course, without critical comment by those whose purpose is to establish superior-inferior relationships among races of man. They have been cited reluctantly by those who appear to wish that facts were otherwise but who see no escape from scientific data. It is to this explosive issue that I now turn. The main point of my essay, however, is not whether blacks are genetically inferior to whites as many persons have claimed on the basis of IQ scores; this is a point that can be shunted aside momentarily. The question I raise here is whether the available data really bear at all on the genetics of the relative intelligence of different races. In order to raise this question properly, it will be necessary to say something about the design of scientific experiments.

What is meant by the design of an experiment? When is an experiment well designed? An experiment is more than a gathering of data or a recording of observations. A well-designed experiment manipulates events so that the data it yields are relevant to a specific question. All experiments ask questions; a well-designed experiment asks the question that the experimenter intends to ask.

The analysis of experimental results in the case of a well-planned experiment proceeds according to a predetermined plan; that is, the analysis of the data and the identification of the conclusions they support are included in the original design. Ad hoc, last-minute interpretations and a posteriori rationalizations are not part of a good experiment.

Abandoning results obtained from a well-designed experiment or refusing to accept the conclusion to which they lead is to a thoughtful experimenter a serious act, one that compels him to undertake a thorough review of the questions he is attempting to ask and the procedures he is using in asking them. An experimental technique that is used routinely only as long as the results it yields agree with preconceived notions, but whose results are discarded or otherwise ignored when they deviated from those expected, is a worthless technique. The data accumulated through its use, as any experimentalist knows, are useless for reaching decisions.

The measurement of the IQ of persons belonging to different races has, with minor variations followed a standard format. The examiner enters a classroom and gives the students an examination; the procedural details as well as the nature of the test may vary considerably from study to study but in the end the students have been examined so that each can be assigned a test score. The examiner (or others associated with him) classifies the students (subjects) as Negro and white; Negro, intermediate, and white; or according to an even finer scale. The examination scores are then analyzed according to the assigned racial classification of the students. More refined analyses might introduce correction factors for family income, family size, age, after-school employment, or any of a large number of such items. The one item for which a correction factor is not employed, of course, is "racial characteristics" because the aim of the study is to determine the relationship — ostensibly the *genetic* relationship — between these and IQ.

Hundreds of these studies have been made. The great majority of them agree in showing that the IQ scores of Negroes is several points below that of whites. When the results of an occasional study differ from the usual and by now expected pattern, rationalizations are offered in explanation. These explanations may or may not be relevant since their need seems never to have been anticipated by the investigator nor their validity subsequently confirmed. The sheer bulk of the results obtained by tests such as these has driven some persons to conclude that blacks *must* differ genetically from whites in intelligence.

Dr. A. M. Shuey, who has compiled one of the largest review of IQ tests, concludes:*

> The remarkable consistency in test results, whether they pertain to school or preschool children, to children between Ages 6 to 9 or 10 to 12, to children in Grades 1 to 3 or 4 to 7, to high school or college students, to enlisted men or officers in training in the Armed Forces – in World War I, World War II, or the Post-Korean period – to veterans of the Armed Forces, to homeless men or transients, to gifted or mentally deficient, to deliquent or criminal . . . point[s] to the presence of native differences between Negroes and whites as determined by intelligence tests.

The testing procedure just described has nothing whatsoever to do with genetics. Neither the study of students in one classroom nor the compilation of hundreds of studies involving hundreds of classrooms can make the observations adequate to support the claim that genes or chromosomal segments of African origin are less adapted for the advanced intellectual development of an individual than are those of European origin.

From the genetics laboratories of universities throughout the world, I can obtain strains of vinegar flies known to be extremely sensitive to carbon dioxide. These strains breed true. They would exhibit their sensitivity to carbon dioxide, I am sure, at both the North and South poles as well as at the equator, at sea level or on the highest mountains, at various humidities, or under a wide variety of other experimental conditions. Following Shuey's example, I could also write: "The remarkable consistency of these results . . . point to the presence of some native differences" between normal and carbon-dioxide-sensitive flies. If I were to follow her example, my conclusion would be wrong. Sensitivity to carbon dioxide is not a genetically caused trait; it is environmental. An infection of flies by a virus-like organism is responsible for the abnormal trait; any strain of vinegar flies can be made sensitive to carbon dioxide by injecting the virus particles into the parental flies through a fine hypodermic needle.

The claim that systematic differences in IQ scores of blacks and whites is evidence for a genetic difference between the two groups is simply unsubstantiated. This bald statement is made here quite independently of the objections of both those who distrust the concept of an "intelligence quotient" and those who deny the existence of races. A student of genetics who approached his professor with data on vinegar flies comparable to those gathered by educational psychologists and who claimed that his data constituted evidence for a genetic difference between strains would be sent back to his laboratory bench to gather data *of a different sort*. He would be told to make crosses between strains and to return with *genetic* evidence, evidence that the trait involved could be associated with segregating genes.

Genetic differences between strains are not proven by systematic differences between strains; they are proven by the parallel inheritance of these

*Audrey M. Shuey, *The Testing of Negro Intelligence*, 2nd ed. (New York: Social Science Press, 1966), p. 520.

differences and of segregating genetic material; they are proven by the recombination of traits contributed by two parents. The *genetics* of bacteria, despite early papers of which at least one is now regarded as a classic, was nearly as much a matter of faith as of science until 1946 when Drs. Joshua Lederberg and E.L. Tatum reported on gene recombination in the colon bacterium. The differences in IQ that one can demonstrate between blacks and whites are of no use in discussing possible genetic differences between races in respect to IQ scores. They show only that under present cultural conditions those persons who are regarded as Negroes do more poorly on certain types of examinations than do those persons who are classified white.

Are there experimental tests that might yield data bearing on interracial genetic differences in respect to intelligence? Yes, such tests could be devised; they would rely heavily upon alternative forms of genes with no visible external effects for which Europeans and African Negroes differ markedly. They would be expensive tests to carry out. Furthermore, it is doubtful whether those tests that are feasible would completely eliminate the effects of cultural heredity − of the nongenetic influences of parents on their children.

There is a serious question, too, as to whether such tests are worthwhile. They certainly are not required to settle the matter of a possible racial average that an individual white or black might carry unseen because there is no such thing. They are not to be made in order to change the minds of those persons who are emotionally committed to the idea of white supremacy because experimental evidence does not alter the beliefs of the emotionally committed. Is there any reason, therefore, for carrying out a well-designed experiment on the effect of racial background on IQ? Only that proposed by a young student: "The techniques and results would be on record so that when the old persons vanish, the young will be able to study the data and to see what they mean." But this student was too young to understand the cost in both dollars and time that a well-designed experiment in this field of research would demand. Whether it is worth the great effort is, in my opinion, still a moot question.

SECTION FOUR

Radiation Biology

Introduction

The audience was young. The boys and their girls had been polite although a few had glanced at each other and squirmed with impatience. The Thursday-evening guest lecture was a traditional ordeal for these fraternity men and their dates; tonight's lecture had been neither worse nor better than many earlier ones. Finally, that's it! Any questions? Questions? Yes?

"When we exploded that atomic bomb in the Philippines or wherever it was . . .?"

The audience was young indeed. They didn't remember Pearl Harbor the way I did. It was December 7, 1941. Three of us were in the zoology laboratory that Sunday afternoon; we were making slides for Professor Schrader's cytology course. A postdoctoral fellow was across the hall studying the interaction of mutant genes during the development of the vinegar fly. December 7, 1941 – it was the beginning of a long hiatus in the lives of most of today's middle-aged Americans. The young man who asked the question at the fraternity lecture was not yet born.

Where is Hiroshima? Where is Nagasaki? I remember the days of their destruction – August, 1945. I was stationed at Maxwell Field, Montgomery, Alabama. An overpass ran from the barracks area to the athletic field; beneath it was the highway – Route 143, perhaps; it made no difference then, and it is impossible to tell by today's road map – on which automobiles took civilians where they wanted to go, when they wanted to go. Or so it seemed to the soldiers on the base. On the field every move outside the barracks – to class, to mess, to the Post Exchange – was in marching formation only. Then, the bomb was dropped; next the war was over. During the following weeks small formations of soldiers, discharge papers in hand, stood at attention in ranks for the last time, listening to the strains of "Going Home." Over and over at half-hour intervals, for each new group as it assembled, the base band played the same sad song – "Going Home." The fraternity boy was about three at the time; his father may well have come home that week, or that month, *because* the bombs were dropped – no, not on the Philippines, but on Japan – first on Hiroshima and then on Nagasaki. We were at war with Japan. Atomic bombs were used, not for purposes more evil than war itself, but to allow those in uniform to survive and to return home at last.

Radiation biology thrives largely because of atomic and nuclear bombs. It is a branch of biology founded on paradoxes. That radiation could cause cancer was discovered by the very earliest radiologists; that it could cure cancer was also

171

discovered almost at once. The causes for concern over radiation hazards were well known for decades preceding the development and detonation of the first atomic bomb. Despite this knowledge, a number of the young atomic physicists of the 1940's have since died of leukemia. The mushrooming interest in radiation biology following World War II was based in large part on the possibility that by prompt treatment, the casualties of an all-out atomic war might be minimized. The development of large nuclear devices (H-bombs) has made the idea of survivors and their sophisticated medical treatment largely meaningless. First, you have to have survivors.

Hiroshima, the first selection chosen to introduce this section, has captured one month in the lives of Mrs. Nakamura and her children — a month that begins at her kitchen door 1350 yards from the center of the first A-bomb explosion and ends at the home of her sister-in-law where she suffers the debilitation after effects of radiation exposure.

Toshio Nakamura, ten years old at the time of the explosion, helped two small girl friends look for their mothers; later he was to write in a grammar-school composition, "Kikuki's mother was wounded and Murakami's mother, alas, was dead." I do not know what has happened to Mrs. Nakamura and her children Toshio, Yaeko, and Myeko since August, 1945. The sorts of things that the radiation exposure did to Mrs. Nakamura and her children, or may do to her grandchildren are topics for the essays of this section.

A second selection has been included in this section; it is a chapter entitled "Atoms and Radiations" from a book, *Radiation, Genes, and Man,* by Professor Th. Dobzhansky and myself. This selection is not great literature and for that I should apologize. I am faced, however, with a dilemma. An understanding of radiation biology — an understanding of Mrs. Nakamura's ordeal — requires some knowledge of atomic structure and radiation physics. Neither of these topics lends itself to inspired composition. Were I to attempt to include the necessary information in an essay, I would unavoidably plod through technical details saying much that Dobzhansky and I have said already. Consequently, the chapter from *Radiation, Genes, and Man* is presented (slightly modified) as a literary selection that provides a limited (but necessary) background in radiation physics.

Hiroshima*

John Hersey

At nearly midnight, the night before the bomb was dropped, an announcer on the city's radio station said that about two hundred B-29s were approaching southern Honshu and advised the population of Hiroshima to evacuate to their designated "safe areas." Mrs. Hatsuyo Nakamura, the tailor's widow, who lived in the section called Nobori-cho and who had long had a habit of doing as she was told, got her three children — a ten-year-old boy, Toshio, an eight-year-old girl, Yaeko, and a five-year-old girl, Myeko — out of bed and dressed them and walked with them to the military area known as the East Parade Ground, on the northeast edge of the city. There she unrolled some mats and the children lay down on them. They slept until about two, when they were awakened by the roar of the planes going over Hiroshima.

As soon as the planes had passed, Mrs. Nakamura started back with her children. They reached home a little after two-thirty and she immediately turned on the radio, which, to her distress, was just then broadcasting a fresh warning. When she looked at the children and saw how tired they were, and when she thought of the number of trips they had made in past weeks, all to no purpose, to the East Parade Ground, she decided that in spite of the instructions on the radio, she simply could not face starting out all over again. She put the children in their bedrolls on the floor, lay down herself at three o'clock, and fell asleep at once, so soundly that when planes passed over later, she did not waken to their sound.

The siren jarred her awake at about seven. She arose, dressed quickly, and hurried to the house of Mr. Nakamoto, the head of her Neighborhood Association, and asked him what she should do. He said that she should remain at home unless an urgent warning — a series of intermittent blasts of the siren — was sounded. She returned home, lit the stove in the kitchen, set some rice to cook, and sat down to read that morning's Hiroshima Chugoku. To her relief, the all-clear sounded at eight o'clock. She heard the children stirring, so she went and gave each of them a handful of peanuts and told them to stay on their bedrolls, because they were tired from the night's walk. She had hoped that they would go back to sleep, but the man in the house directly to the south began to make a terrible hullabaloo of hammering, wedging, ripping, and

---

splitting. The prefectural government, convinced, as everyone in Hiroshima was, that the city would be attacked soon, had begun to press with threats and warnings for the completion of wide fire lanes, which, it was hoped, might act in conjunction with the rivers to localize any fires started by an incendiary raid; and the neighbor was reluctantly sacrificing his home to the city's safety. Just the day before, the prefecture had ordered all able-bodied girls from the secondary schools to spend a few days helping to clear these lanes, and they started work soon after the all-clear sounded.

Mrs. Nakamura went back to the kitchen, looked at the rice, and began watching the man next door. At first, she was annoyed with him for making so much noise, but then she was moved almost to tears by pity. Her emotion was specifically directed toward her neighbor, tearing down his home, board by board, at a time when there was so much unavoidable destruction, but undoubtedly she also felt a generalized, community pity, to say nothing of self-pity. She had not had an easy time. Her husband, Isawa, had gone into the Army just after Myeko was born, and she had heard nothing from or of him for a long time, until, on March 5, 1942, she received a seven-word telegram: "Isawa died an honorable death at Singapore." She learned later that he had died on February 15th, the day Singapore fell, and that he had been a corporal. Isawa had been a not particularly prosperous tailor, and his only capital was a Sankoku sewing machine. After his death, when his allotments stopped coming, Mrs. Nakamura got out the machine and began to take in piecework herself, and since then had supported the children, but poorly, by sewing.

As Mrs. Nakamura stood watching her neighbor, everything flashed whiter than any white she had ever seen. She did not notice what happened to the man next door; the reflex of a mother set her in motion toward her children. She had taken a single step (the house was 1,350 yards, or three-quarters of a mile, from the center of the explosion) when something picked her up and she seemed to fly into the next room over the raised sleeping platform, pursued by parts of her house.

Timbers fell around her as she landed, and a shower of tiles pommelled her; everything became dark, for she was buried. The debris did not cover her deeply. She rose up and freed herself. She heard a child cry, "Mother, help me!," and saw her youngest — Myeko, the five-year-old — buried up to her breast and unable to move. As Mrs. Nakamura started frantically to claw her way toward the baby, she could see or hear nothing of her other children.

●   ●   ●

Mrs. Hatsuyo Nakamura, the tailor's widow, having struggled up from under the ruins of her house after the explosion, and seeing Myeko, the youngest of her three children, buried breast-deep and unable to move, crawled across the debris, hauled at timbers and flung tiles aside, in a hurried effort to free the child. Then, from what seemed to be caverns far below, she heard two small voices crying, "Tasukete! Tasukete! Help! Help!"

She called the names of her ten-year-old son and eight-year-old daughter: "Toshio! Yaeko!"

The voices from below answered.

Mrs. Nakamura abandoned Myeko, who at least could breathe, and in a frenzy made the wreckage fly above the crying voices. The children had been sleeping nearly ten feet apart, but now their voices seemed to come from the same place. Toshio, the boy, apparently had some freedom to move, because she could feel him undermining the pile of wood and tiles as she worked from above. At last she saw his head, and she hastily pulled him out by it. A mosquito net was wound intricately, as if it had been carefully wrapped, around his feet. He said he had been blown right across the room and had been on top of his sister Yaeko under the wreckage. She now said, from underneath, that she could not move, because there was something on her legs. With a bit more digging, Mrs. Nakamura cleared a hole above the child and began to pull her arm. "Itai! It hurts!" Yaeko cried. Mrs. Nakamura shouted, "There's no time now to say whether it hurts or not," and yanked her whimpering daughter up. Then she freed Myeko. The children were filthy and bruised, but none of them had a single cut or scratch.

Mrs. Nakamura took the children out into the street. They had nothing on but underpants, and although the day was very hot, she worried rather confusedly about their being cold, so she went back into the wreckage and burrowed underneath and found a bundle of clothes she had packed for an emergency, and she dressed them in pants, blouses, shoes, padded-cotton air-raid helmets called bokuzuki, and even, irrationally, overcoats. The children were silent, except for the five-year-old, Myeko, who kept asking questions: "Why is it night already? Why did our house fall down? What happened?" Mrs. Nakamura, who did not know what had happened (had not the all-clear sounded?), looked around and saw through the darkness that all the houses in her neighborhood had collapsed. The house next door, which its owner had been tearing down to make way for a fire lane, was now very thoroughly, if crudely, torn down; its owner, who had been sacrificing his home for the community's safety, lay dead. Mrs. Nakamoto, wife of the head of the local air-raid-defense Neighborhood Association, came across the street with her head all bloody, and said that her baby was badly cut; did Mrs. Nakamura have any bandage? Mrs. Nakamura did not, but she crawled into the remains of her house again and pulled out some white cloth that she had been using in her work as a seamstress, ripped it into strips, and gave it to Mrs. Nakamoto. While fetching the cloth, she noticed her sewing machine; she went back in for it and dragged it out. Obviously, she could not carry it with her, so she unthinkingly plunged her symbol of livelihood into the receptacle which for weeks had been her symbol of safety — the cement tank of water in front of her house, of the type every household had been ordered to construct against a possible fire raid.

A nervous neighbor, Mrs. Hataya, called to Mrs. Nakamura to run away with her to the woods in Asano Park — an estate, by the Kyo River not far off, belonging to the wealthy Asano family, who once owned the Toyo Kisen Kaisha steamship line. The park had been designated as an evacuation area for their neighborhood. Seeing fire breaking out in a nearby ruin (except at the very center, where the bomb itself ignited some fires, most of Hiroshima's citywide conflagration was caused by inflammable wreckage falling on cookstoves and live wires), Mrs. Nakamura suggested going over to fight it. Mrs. Hataya said, "Don't be foolish. What if planes come and drop more bombs?" So Mrs. Nakamura started out for Asano Park with her children and Mrs. Hataya, and she carried her rucksack of emergency clothing, a blanket, an umbrella, and a suitcase of things she had cached in her air-raid shelter. Under many ruins, as they hurried along, they heard muffled screams for help. The only building they saw standing on their way to Asano Park was the Jesuit mission house, alongside the Catholic kindergarten to which Mrs. Nakamura had sent Myeko for a time. As they passed it, she saw Father Kleinsorge, in bloody underwear, running out of the house with a small suitcase in his hand.

● ● ●

All day, people poured into Asano Park. This private estate was far enough away from the explosion so that its bamboos, pines, laurel, and maples were still alive, and the green place invited refugees — partly because they believed that if the Americans came back, they would bomb only buildings; partly because the foliage seemed a center of coolness and life, and the estate's exquisitely precise rock gardens, with their quiet pools and arching bridges, were very Japanese, normal, secure; and also partly (according to some who were there) because of an irresistible, atavistic urge to hide under leaves. Mrs. Nakamura and her children were among the first to arrive, and they settled in the bamboo grove near the river. They all felt terribly thirsty, and they drank from the river. At once they were nauseated and began vomiting, and they retched the whole day. Others were also nauseated; they all thought (probably because of the strong order of ionization, an "electric smell" given off by the bomb's fission) that they were sick from a gas the Americans had dropped. When Father Kleinsorge and the other priests came into the park, nodding to their friends as they passed, the Nakamuras were all sick and prostrate. A woman named Iwasaki, who lived in the neighborhood of the mission and who was sitting near the Nakamuras, got up and asked the priests if she should stay where she was or go with them. Father Kleinsorge said, "I hardly know where the safest place is." She stayed there, and later in the day, though she had no visible wounds or burns, she died.

● ● ●

The roar of approaching planes was heard about this time. Someone in the crowd near the Nakamura family shouted, "It's some Grummans coming to strafe us!" A baker named Nakashima stood up and commanded, "Everyone who is wearing anything white, take it off." Mrs. Nakamura took the blouses off her children, and opened her umbrella and made them get under it. A great number of people, even badly burned ones, crawled into bushes and stayed there until the hum, evidently of a reconnaissance or weather run, died away.

It began to rain. Mrs. Nakamura kept her children under the umbrella. The drops grew abnormally large, and someone shouted, "The Americans are dropping gasoline. They're going to set fire to us!" (This alarm stemmed from one of the theories being passed through the park as to why so much of Hiroshima had burned: it was that a single plane had sprayed gasoline on the city and they somehow set fire to it in one flashing moment.) But the drops were palpably water, and as they fell, the wind grew stronger and stronger, and suddenly — probably because of the tremendous convection set up by the blazing city — a whirlwind ripped through the park. Huge trees crashed down; small ones were uprooted and flew into the air. Higher, a wild array of flat things revolved in the twisting funnel — pieces of iron roofing, papers, doors, strips of matting. Father Kleinsorge put a piece of cloth over Father Schiffer's eyes, so that the feeble man would not think he was going crazy. The gale blew Mrs. Murata, the mission housekeeper, who was sitting close by the river, down the embankment at a shallow, rocky place, and she came out with her bare feet bloody. The vortex moved out onto the river, where it sucked up a waterspout and eventually spent itself.

• • •

In the park, Mrs. Murata kept Father Kleinsorge awake all night by talking to him. None of the Nakamura family were able to sleep, either; the children, in spite of being very sick, were interested in everything that happened. They were delighted when one of the city's gas-storage tanks went up in a tremendous burst of flame. Toshio, the boy, shouted to the others to look at the reflection in the river. Mr. Tanimoto, after his long run and his many hours of rescue work, dozed uneasily. When he awoke, in the first light of dawn, he looked across the river and saw that he had not carried the festered, limp bodies high enough on the sandspit the night before. The tide had risen above where he had put them; they had not had the strength to move; they must have drowned. He saw a number of bodies floating in the river.

• • •

The Jesuits took about fifty refugees into the exquisite chapel of the Novitiate. The rector gave them what medical care he could — mostly just the cleaning away of pus. Each of the Nakamuras was provided with a blanket and a mosquito net. Mrs. Nakamura and her younger daughter had no appetite and ate nothing; her son and other daughter ate, and lost, each meal they were offered. On August 10th, a friend, Mrs. Osaki, came to see them and told them that her son Hideo had been burned alive in the factory where he worked. This Hideo had been a kind of hero to Toshio, who had often gone to the plant to watch him run his machine. That night, Toshio woke up screaming. He had dreamed that he had seen Mrs. Osaki coming out of an opening in the ground with her family, and then he saw Hideo at his machine, a big one with a revolving belt, and he himself was standing beside Hideo, and for some reason this was terrifying.

●   ●   ●

About a week after the bomb dropped, a vague, incomprehensible rumor reached Hiroshima — that the city had been destroyed by the energy released when atoms were somehow split in two. The weapon was referred to in this word-of-mouth report as genshi bakudan — the root characters of which can be translated as "original child bomb." No one understood the idea or put any more credence in it than in the powdered magnesium and such things. Newspapers were being brought in from other cities, but they were still confining themselves to extremely general statements, such as Domei's assertion on August 12th: "There is nothing to do but admit the tremendous power of this inhuman bomb." Already, Japanese physicists had entered the city with Lauritsen electroscopes and Neher electrometers; they understood the idea all too well.

On August 12th, the Nakamuras, all of them still rather sick, went to the nearby town of Kabe and moved in with Mrs. Nakamura's sister-in-law. The next day, Mrs. Nakamura, although she was too ill to walk much, returned to Hiroshima alone, by electric car to the outskirts, by foot from there. All week, at the Novitiate, she had worried about her mother, brother, and older sister, who had lived in the part of town called Fukuro, and besides, she felt drawn by some fascination, just as Father Kleinsorge had been. She discovered that her family were all dead. She went back to Kabe so amazed and depressed by what she had seen and learned in the city that she could not speak that evening.

●   ●   ●

In Kabe, on the morning of August 15th, ten-year-old Toshio Nakamura heard an airplane overhead. He ran outdoors and identified it with a professional eye as a B-29. "There goes Mr. B!" he shouted.

One of his relatives called out to him, "Haven't you had enough of Mr. B?"

The question had a kind of symbolism. At almost that very moment, the dull, dispirited voice of Hirohito, the Emperor Tenno, was speaking for the first time in history over the radio: "After pondering deeply the general trends of the world and the actual conditions obtaining in Our Empire today, We have decided to effect a settlement of the present situation by resorting to an extraordinary measure. . . ."

Mrs. Nakamura had gone to the city again, to dig up some rice she had buried in her Neighborhood Association air-raid shelter. She got it and started back for Kabe. On the electric car, quite by chance, she ran into her younger sister, who had not been in Hiroshima the day of the bombing. "Have you heard the news?" her sister asked.

"What news?"

"The war is over."

"Don't say such a foolish thing, sister."

"But I heard it over the radio myself." And then, in a whisper, "It was the Emperor's voice."

"Oh," Mrs. Nakamura said (she needed nothing more to make her give up thinking, in spite of the atomic bomb, that Japan still had a chance to win the war), "in that case . . ."

•   •   •

As she dressed on the morning of August 20th, in the home of her sister-in-law in Kabe, not far from Nagatsuka, Mrs. Nakamura, who had suffered no cuts or burns at all, though she had been rather nauseated all through the week she and her children had spent as guests of Father Kleinsorge and the other Catholics at the Novitiate, began fixing her hair and noticed, after one stroke, that her comb carried with it a whole handful of hair; the second time, the same thing happened, so she stopped combing at once. But in the next three or four days, her hair kept falling out of its own accord, until she was quite bald. She began living indoors, practically in hiding. On August 26th, both she and her younger daughter, Myeko, woke up feeling extremely weak and tired, and they stayed on their bedrolls. Her son and other daughter, who had shared every experience with her during and after the bombing, felt fine.

•   •   •

Because so many people were suddenly feeling sick nearly a month after the atomic bomb was dropped, an unpleasant rumor began to move around, and eventually it made its way to the house in Kabe where Mrs. Nakamura lay bald and ill. It was that the atomic bomb had deposited some sort of poison on Hiroshima which would give off deadly emanations for seven years; nobody could go there all that time. This especially upset Mrs. Nakamura, who remembered that in a moment of confusion on the morning of the explosion she had literally sunk her entire means of livelihood, her Sankoku sewing machine, in the small cement water tank in front of what was left of her house; now no one would be able to go and fish it out. Up to this time, Mrs. Nakamura and her relatives had been quite resigned and passive about the moral issue of the atomic bomb, but this rumor suddenly aroused them to more hatred and resentment of America than they had felt all through the war.

Japanese physicists, who knew a great deal about atomic fission (one of them owned a cyclotron), worried about lingering radiation at Hiroshima, and in mid-August, not many days after President Truman's disclosure of the type of bomb that had been dropped, they entered the city to make investigations. The first thing they did was roughly to determine a center by observing the side on which telephone poles all around the heart of the town were scorched; they settled on the torii gateway of the Gokoku Shrine, right next to the parade ground of the Chugoku Regional Army Headquarters. From there, they worked north and south with Lauritsen electroscopes, which are sensitive to both beta particles and gamma rays. These indicated that the highest intensity of radioactivity, near the torii, was 4.2 times the average natural "leak" of ultra-short waves for the earth of that area. The scientists noticed that the flash of the bomb had discolored concrete to a light reddish tint, had scaled off the surface of granite, and had scorched certain other types of building material, and that consequently the bomb had, in some places, left prints of the shadows that had been cast by its light. The experts found, for instance, a permanent shadow thrown on the roof of the Chamber of Commerce Building (220 yards from the rough center) by the structure's rectangular tower; several others in the lookout post on top of the Hypothec Bank (2,050 yards); another in the tower of the Chugoku Electric Supply Building (800 yards); another projected by the handle of a gas pump (2,630 yards); and several on granite tombstones in the Gokoku Shrine (385 yards). By triangulating these and other such shadows with the objects that formed them, the scientists determined that the exact center was a spot a hundred and fifty yards south of the torii and a few yards southeast of the pile of ruins that had once been the Shima Hospital. (A few vague human silhouettes were found, and these gave rise to stories that eventually included fancy and precise details. One story told how a painter on a ladder was monumentalized in a kind of bas-relief on the stone facade of a bank building on which he was at work, in the act of dipping his brush into his paint can; another, how a man and his cart on the bridge near the Museum of Science and Industry, almost under the center of the explosion, were cast down in an embossed shadow which made it clear that the man was about to whip his horse.) Starting east and west from the actual center, the scientists, in early September, made new measurements, and the highest radiation they found this time was 3.9 times the natural "leak." Since radiation of at least a thousand times the natural "leak" would be required to cause serious effects on the human body, the scientists announced that people could enter Hiroshima without any peril at all.

As soon as this reassurance reached the household in which Mrs. Nakamura was concealing herself — or, at any rate, within a short time, after her hair had started growing back again — her whole family relaxed their extreme hatred of America, and Mrs. Nakamura sent her brother-in-law to look for the sewing machine. It was still submerged in the water tank, and when he brought it home, she saw, to her dismay, that it was all rusted and useless.

· · ·

Mrs. Nakamura lay indoors with Myeko. They both continued sick, and though Mrs. Nakamura vaguely sensed that their trouble was caused by the bomb, she was too poor to see a doctor and so never knew exactly what the matter was. Without any treatment at all, but merely resting, they began gradually to feel better. Some of Myeko's hair fell out, and she had a tiny burn on her arm which took months to heal. The boy, Toshio, and the older girl, Yaeko, seemed well enough, though they, too, lost some hair and occasionally had bad headaches. Toshio was still having nightmares, always about the nineteen-year-old mechanic, Hideo Osaki, his hero, who had been killed by the bomb.

•   •   •

It came to Mrs. Nakamura's attention that a carpenter from Kabe was building a number of wooden shanties in Hiroshima which he rented for fifty yen a month — $3.33, at the fixed rate of exchange. Mrs. Nakamura had lost the certificates of her bonds and other wartime savings, but fortunately she had copied off all the numbers just a few days before the bombing and had taken the list to Kabe, and so, when her hair had grown in enough for her to be presentable, she went to her bank in Hiroshima, and a clerk there told her that after checking her numbers against the records the bank would give her her money. As soon as she got it, she rented one of the carpenter's shacks. It was in Nobori-cho, near the site of her former house, and though its floor was dirt and it was dark inside, it was at least a home in Hiroshima, and she was no longer dependent on the charity of her in-laws. During the spring, she cleared away some nearby wreckage and planted a vegetable garden. She cooked with utensils and ate off plates she scavenged from the debris. She sent Myeko to the kindergarten which the Jesuits reopened, and the two older children attended Nobori-cho Primary School, which, for want of buildings, held classes out of doors. Toshio wanted to study to be a mechanic like his hero, Hideo Osaki. Prices were high; by midsummer Mrs. Nakamura's savings were gone. She sold some of her clothes to get food. She had once had several expensive kimonos, but during the war one had been stolen, she had given one to a sister who had been bombed out in Tokuyama, she had lost a couple in the Hiroshima bombing, and now she sold her last one. It brought only a hundred yen, which did not last long. In June, she went to Father Kleinsorge for advice about how to get along, and in early August, she was still considering the two alternatives he suggested — taking work as a domestic for some of the Allied occupation forces, or borrowing from her relatives enough money, about five hundred yen, or a bit more than thirty dollars, to repair her rusty sewing machine and resume the work of a seamstress.

•   •   •

It would be impossible to say what horrors were embedded in the minds of the children who lived through the day of the bombing in Hiroshima. On the surface, their recollections, months after the disaster, were of an exhilarating adventure. Toshio Nakamura, who was ten at the time of the bombing, was soon able to talk freely, even gaily, about the experience, and a few weeks before the anniversary he wrote the following matter-of-fact essay for his teacher at Nobori-cho Primary School: "The day before the bomb, I went for a swim. In the morning, I was eating peanuts. I saw a light. I was knocked to little sister's sleeping place. When we were saved, I could only see as far as the tram. My mother and I started to pack our things. The neighbors were walking around burned and bleeding. Hataya-san told me to run away with her. I said I wanted to wait for my mother. We went to the park. A whirlwind came. At night a gas tank burned and I saw the reflection in the river. We stayed in the park one night. Next day I went to Taiko Bridge and met my girl friends Kikuki and Murakami. They were looking for their mothers. But Kikuki's mother was wounded and Murakami's mother, alas, was dead."

Atoms and Radiation*

Bruce Wallace
Th. Dobzhansky

Life and Radiation

The sun and its life-giving rays have inspired wonder and awe in men of every age; sun worship was a part of many religions. Here are some lines from the HYMN TO THE SUN composed by Ikhnaton, a pharaoh of Egypt, more than three thousand years ago:

> Thou art he who createst the man-child in women,
> Who makest seed in man,
> Who giveth life to the son in the body of his mother,
> Who soothest him that he may not weep,
> A nurse even in the womb,
> Who giveth breath to animate every one that he maketh.

Ikhnaton was right; life endures on earth because living matter is adapted to capture and to store within itself some of the solar energy that reaches the earth. Life has existed on earth for perhaps two billion years; it has learned to carry on, literally and metaphorically, "under the sun." Until man started to use the energy of atomic fission, every bit of energy he consumed came ultimately from solar radiation.

But there exist other kinds of radiations in nature which are weak or absent in the sun's rays when the latter reach the earth's surface. Some of these radiations are destructive to biological organization. Outstanding among these destructive agents are the radiations variously called ionizing, penetrating, short-wave, or high-energy radiations, which are produced by radium and other radioactive substances. Man has learned to create them artificially; some of them are generated in the processes of atomic fission and fusion.

It would be out of place to discuss here the physical nature of radiations in detail. Such discussions can be found in a large number of books, written on different levels of physical and mathematical sophistication. We should, however, be familiar with some facts concerning different kinds of radiations, their measurement, and the amounts man encounters in the natural environment, and in the environments created by our modern industrial civilization.

Molecules and Atoms

The world about us is composed of countless types of substances, some solid, others liquid, still others gaseous. Numerous words in our vocabulary are names of these different substances — water, alcohol, kerosene, salt, sugar, soda, aspirin, etc. We recognize these substances by their consistency, color, smell, taste; by their effects on ourselves or on things or processes in our environment. These properties and effects depend largely on the molecular structure of the substances concerned. If we take a lump of sugar and divide it into smaller and smaller pieces, we still have sugar left after each division until we arrive at very small units called molecules. If a molecule were to be magnified about 100 million times, it could be conveniently represented by a picture on a page. If a lump of sugar weighing one pound were to be magnified 100 million times, it would be almost as large as the earth.

Molecules can be broken into still smaller parts. For example, if we heat sugar we get two quite different substances — namely a black residue of carbon and — although we may not see it as it boils off — some water. The water given off when sugar is heated is exactly the same as that coming from the faucet above the sink; and the carbon is exactly the same as that found in a piece of charcoal or in soot. Water can, in turn, be broken down into two further substances. If an electric current is passed through water, two gases — hydrogen and oxygen — are produced. Again, the oxygen and hydrogen we get from water are precisely the same as those obtained from air or other substances.

What we are left with when we break apart different sorts of molecules is a relatively small number of elements. There exist many thousands, perhaps millions, of different kinds of molecules, but they are made up of only about 92 naturally occurring elements, and even some of these are quite rare in nature. How does an element differ from a molecular compound such as sugar or water? As we have seen, sugar and water can be decomposed into simpler substances quite easily. But even if an elementary substance, such as carbon or oxygen or iron, is divided again and again, it still remains carbon, oxygen, or iron. The ultimate particle of an element, the atom, defies subdivision by ordinary maltreatment. A molecule of sugar is composed of 6 atoms of the element carbon, 6 atoms of the element oxygen, and 12 atoms of the element hydrogen. The properties of sugar are determined by the number and relative position of these atoms within the molecule. Molecules can be broken into atoms by heat, by electric currents, or by reactions with other substances. Atoms cannot be broken by such manipulations. The word atom means "indivisible" in Greek.

The picture of the world composed of a fairly small number of indivisible and immutable atoms has the virtue of great simplicity and clarity. Moreover, this picture is adequate for most purposes of practical life and even of rather advanced technology. During the second half of the nineteenth century and the early years of the current one, most physicists and chemists thought that atoms were, indeed, the ultimate reality of the material world. That this conception was erroneous became clear through the discovery, by Pierre and Marie Curie, of an extraordinary element called radium. This amazing substance was found to emit, continuously and without any outside source of energy, powerful and penetrating radiations, and to transform itself, slowly but inexorably, into a quite different element — namely, lead. Atoms thus proved to have a complex internal structure, and atoms of one element were shown to be capable of turning into those of another.

Anatomy of the Atom:
Radiations

The atoms of the 92 elements known before 1940, and a few more discovered since then, can be regarded as consisting of three basic particles — electrons, protons and neutrons. These particles are extremely small — 10,000 times smaller in diameter than the atom itself. Moreover, the interior of the atom resembles, of all things, the solar system. In the center there is the atomic nucleus, compounded of protons and neutrons; the nucleus may be likened to the sun in the solar system. In the space around the nucleus, and spinning around it like the planets around the sun, are a number of electrons. The number of orbital electrons is characteristic for atoms of each element.

The simplest atom is that of the element hydrogen. A hydrogen atom has a diameter of about 0.000,000,005 of an inch. The center of the atom is occupied by the nucleus; the nucleus contains but a single proton, which is an electrically charged particle positive in sign. Rotating around the nucleus is a single electron. The electron is also electrically charged, but its charge is negative, thus balancing the positive charge of the nucleus. The mass of the electron is, however, very tiny — about $1/1,840$ of that of the proton. Most of the weight of the atom is thus concentrated in the nucleus. Atoms of other elements are more complex. Their nuclei contain varying numbers of protons and neutrons depending upon the element, and many electrons gyrate around the nuclei in fixed orbits. For each element the number of electrons balances the number of protons in the nucleus. Thus, an atom of one of the heaviest elements, uranium, has a nucleus compounded of 146 neutrons and 92 protons; 92 electrons spin around the nucleus in several concentric "shells." Atoms of other elements range in complexity, and in weight, between those of the lightest element, hydrogen, and up to and beyond uranium at the upper end.

If the properties of electrons are to be studied easily they must be removed from the atoms containing them. One of the early tools used for this purpose by physicists was the Geissler tube, a glass tube from which most of the air has been evacuated. Inside, at each end of the tube, is a metallic plate connected to the poles of a high-voltage source of electricity by a wire that pierces the glass. When the current is turned on the tube gives off a glow, which is caused by the passage of a stream of electrons through the rarefied air. The electrons can be demonstrated to leave the plate connected to the negative pole and to jump across the tube to the plate connected to the positive pole. By coating the positive plate with a phosphorescent substance (zinc sulfide), one is able to discern the impact of the electrons as pinpoints of light.

In 1895 a German physicist, W. K. Roentgen, discovered that hitherto unknown radiations were coming out of Geissler tubes. The properties of these radiations seemed mysterious enough for them to be called x-rays — "x" designating the unknown nature of the rays. X-rays are invisible to the human eye, but they can be detected by photographic plates; strangest of all, x-rays easily penetrate screens of black paper and other materials completely opaque to ordinary rays of light.

The penetrating powers of x-rays made them important in ways which were not obvious to their discoverer and to many others. They became indispensable tools for the diagnosis of many diseases and for the treatment of some of them. Well-equipped hospitals, and even doctors' and dentists' offices, now usually include x-ray machines. Unfortunately, together with properties useful to man, these rays proved to have, when improperly used, some sinister ones as well. They can cause severe damage to the body and, as we have known since 1927, to the units of heredity, the genes.

X-rays are not streams of electrons escaping from the Geissler tube; electrons cannot pass easily through glass or heavy paper. X-rays are a part of the radiation spectrum, of which visible light is one part and radio waves still another. The properties of radiation depend upon the length of its waves. Radio waves vary in length from several kilometers to a few meters; microwaves (radar waves) from meters to millimeters; infrared (heat) waves from about a millimeter to 8/10,000 (8×10^{-4}) of a millimeter; visible light occupies the narrow band from 8×10^{-4} to about 10^{-5} of a millimeter; ultraviolet from 10^{-5} to 10^{-6}; while x-rays extend from some 10^{-6} (soft x-rays) to 10^{-8} (hard x-rays) and to 10^{-9} (known also as gamma rays). Still shorter wavelengths (those characteristic of cosmic rays) are known.

Another basic fact of radiation physics is that radiation behaves as if it comes in discrete packages; these are called quanta. Each quantum of radiation has in it a certain amount of energy; the shorter the wavelength of radiation, the greater the energy contained in its quanta. Thus, a quantum of visible light has relatively little energy, quanta of ultraviolet light are more powerful, and those of x- and gamma rays are more energetic still. This is why x-rays and gamma rays are known as high-energy radiations.

Effects of Radiation on Matter

When visible light falls on a material body, the light quanta are either absorbed in or reflected from the body. The energy carried in the quanta of visible light is, however, not sufficient to affect the structure or behavior of most atoms. Quanta of light resemble peas shot from a toy gun against a wall; the peas bounce off and do not go through the wall. Only some few substances are "phosphorescent." This means that when they are exposed to light, electrons are displaced in their atomic orbits and the substance gives off a glow (that is, emits visible light) for a short time after the exposure stops. The visible radiation is emitted as electrons return to their normal position. Other substances, such as silver bromide, which is used in photographic film, are light-sensitive; their molecules undergo chemical changes because of the energy delivered to them by light.

X-rays, gamma rays, and other high-energy radiations have quanta so powerful that they may perhaps be likened to bullets shot from a rifle. When such a quantum "hits" an atom, its energy is imparted to one of the electrons circling around the atomic nucleus. An "energized" electron may be ejected from an "excited" atom and fly at an enormous speed into surrounding space, where it hits other atoms. Now, an atom that has lost one of its electrons has lost one unit of negative electric charge. It will, therefore, be a positively charged atom, or a positive ion. Such ions are chemically much more active than electrically neutral atoms; they will react easily with other atoms or atomic groups which are able to "share" an electron with them.

The electron ejected from an atom darts through the surrounding space, where it may collide with other electrons in neighboring atoms. These other electrons may in turn be knocked from their orbits. A hit by a powerful x-ray quantum thus produces a whole series of agitated electrons and ionized atoms. The disturbance caused by the initial hit gradually subsides as the energy of the absorbed quantum is distributed among more and more electrons. Finally, no one electron has enough energy left to displace other electrons, and the residue of the energy is dissipated as heat. By then, however, a number of electrons may have been displaced from their atoms; the atoms that have lost electrons are left behind as positive ions; the atoms that have picked up the displaced electrons are negative ions. Each displaced electron results in a pair of ions, one positive and one negative. This is why high-energy radiations are also called ionizing radiations.

Absorption of quanta of high-energy radiation by atoms is not the only cause of ionization. High-speed electrons, called beta rays, exist within x-ray tubes; in fact, x-rays are produced when these electrons strike the metal in the positive electrode. Disintegrating atoms also eject electrons from their orbits at high speed; beta rays are a characteristic emanation of disintegrating radium atoms. These rays have a wide range of velocities and, hence, of characteristic energies. Although more energetic beta rays are known, most of those produced by atomic disintegration are capable of passing through only thin layers of aluminum or about 15 feet of air. They slow down rapidly as they collide with other electrons; their initial energies are rapidly dissipated by ejecting these other electrons from their orbits. Again, the result of irradiation by beta rays is the formation of many pairs of ions.

Since they are simply rapidly moving electrons, beta rays are electrically charged. Another type of charged particle is the alpha particle, a rather ponderous particle bearing two positive electric charges; it is, in fact, the nucleus of a helium atom. Again, velocities of alpha particles vary according to their source; a value of about one twentieth the speed of light is not unusual. For all their weight, alpha particles are not efficient penetrators; a sheet of paper suffices to stop most such particles. However, in living tissue the extent of their penetration is not negligible, especially if their source is inside the body. Throughout the distance they do penetrate, their effect is tremendous; as they speed past and through atoms, they knock electrons out of their orbits at a tremendous rate. Not only electrons but even whole atoms may be dislodged by the blows of these relatively massive particles.

Neutrons have effects of their own. Neutrons, it may be recalled, together with protons make up atomic nuclei. When atomic nuclei disintegrate, neutrons are quite frequently ejected. These may travel at a fast or slow speed. Since they have no electric charge, they can rush or drift right through the electronic orbits of other atoms. Eventually, they are "captured" by the nuclei of these other atoms. For reasons that are not completely known, certain numerical combinations of protons and neutrons are unstable; atoms having such unstable combinations undergo spontaneous nuclear disintegration.

This disintegration, or fission, may result in the ejection of additional neutrons. If, as in the case of uranium-235, the average number of ejected neutrons is larger than 1, there are more free neutrons after the disintegration of each atom than before. It thus becomes possible to initiate and sustain a chain reaction. This reaction may be slow and controlled (as in atomic piles) or almost instantaneous and of great violence (as in atomic, or fission, bombs). Electrons are also frequently expelled from unstable atoms as high-speed beta particles.

Furthermore, the jostling of electrons in their orbits because of nuclear adjustments frequently results in the ejection of a quantum of gamma radiation. Gamma rays are, as we have seen above, very powerful x-rays. Thus, the disintegration of radium atoms results in the release of all three kinds of radiation — alpha, beta, and gamma rays. These rays are all ionizing, and the gamma rays have, in addition, tremendous penetrating powers.

The disintegration of atomic nuclei poses a novel type of biological problem, since the atoms involved change from one chemical element into another. It is obvious that such changes, if they occur in a living cell, may result in grave disturbances. Suppose that a physiological reaction depends upon the presence of sodium atoms; if some of these atoms suddenly change to helium and neon, the normal course of the reaction will be disturbed. The existence of a radioactive isotope of phosphorus poses a special problem for hereditary materials. The deoxyribonucleic acids (DNA), which are the most important constituents of the chromosomes, contain a great deal of phosphorus. If a radioactive variety of phosphorus becomes incorporated into chromosomal DNA, the affected molecule of DNA is doomed. Radioactive phosphorus exists for a few days only, and then changes into sulphur, an element entirely unsuitable for the chemical constitution of DNA.

The normal flow of life processes depends upon the presence, in the proper place and at the proper time, of certain chemical molecules. This is especially true of the processes of heredity. A single change in the DNA of either the sperm or egg cell may make the difference between a normal child and a pitiful invalid. High-energy, or ionizing, radiations disrupt the physical structure of atoms and initiate novel chemical reactions. The addition of a hydrogen atom here, its loss there; the ionization of a formerly uncharged atom; the production of hydrogen peroxide; the physical disruption of long, chainlike molecules — these and a dozen other effects or immediate aftereffects of radiation may raise havoc with the cell's basic organization. In many instances, of course, the damage is done to an expendable substance and is repaired by the cell's built-in defense mechanism. In many other instances, however, damage is done to a unique substance not easily replaced or repaired. Here the effects of radiation are serious. There is no more unique, no more important, no more complex substance than DNA — the chemical that conditions the hereditary processes and directs the manufacture of enzymes and proteins in every cell of the body.

Measurement of Radiation

The effects of radiation upon a living cell or a living body depend upon the amount of radiant energy absorbed. It is obviously important to measure the quantities of radiation applied and to relate them to the biological effects produced. Thus, exposure of the body to high-energy radiations may result in radiation sickness and in death; we must know, then, how much radiation of a given kind will be fatal to a mouse, or to a man, or to some other organism. We must also know how great are the amounts of radiation to which people are exposed from fallout products of testing atomic weapons, from x-rays used in medical practice, and from other sources.

The most useful unit for measuring exposure to x- or gamma rays is the roentgen, usually abbreviated as "r." It is so named to honor the discoverer of x-rays. One roentgen corresponds to the amount of x or gamma radiation which produces about 2×10^9 ion pairs per cubic centimeter of air. This measurement is made fairly easily with the aid of an apparatus (dosimeter) which measures the leakage of electricity from a charged object. The amount of leakage depends on the number of ionized molecules; the greater the amount of x-rays, the greater the number of ions, and the greater the loss of electric charge through leakage.

The number of ions produced by a given amount of radiation is, however, different in different types of substances. One roentgen of x-rays yields, for example, about 1.8×10^{12} ion pairs per gram of living tissue. The measurement of types of radiation other than x- and gamma rays is more involved. The basic unit of the radiation absorbed is the "rad." The relative biological effectiveness of 1 rad of alpha rays is, however, approximately equivalent, under certain conditions, to that of 10 rads of gamma rays. In practice, the biological effects of these radiations are measured by equating them with effects produced by a given dose of x-rays as measured in roentgens. Thus, we have a unit called "rem" (= roentgen-equivalent for man). For measurement of very small amounts of radiation there are "millirads" (mrad, one-thousandth of a rad) and "millirems" (mrem, one-thousandth of a rem).

Sources and Amounts
of Natural Radiation

All life is continuously exposed to high-energy radiations. Such "natural," or backgound, radiations are present everywhere, though their intensity is generally very low. There is every reason to think that such radiations occurred on earth no less abundantly in the past; in fact, we do not know whether the first form of life arose because or in spite of radiation. It would probably be impossible, or at any rate very difficult, to create an environment completely free from such radiations.

Some of the sources of natural radiation are external; others are internal, within living bodies. By far the most omnipresent among the former are cosmic rays. These, as their name suggests, reach the earth from outer space. At the earth's surface cosmic rays consist chiefly of extremely high-energy quanta, corresponding to wavelengths considerably shorter than those of the gamma rays of radium. So great are their penetrating powers that the amounts of this type of radiation received by all organs of the human body are uniform and practically the same outdoors and inside dwellings. These rays have been detected in mines a quarter of a mile or more below the earth's surface. The intensity of cosmic rays varies, however, with the elevation of the locality above sea level; it increases about threefold as one ascends from sea level to 10,000 feet. It also varies with geographic latitude, being generally lower in the tropics than near the magnetic poles of the earth.

Other external sources of background radiation are radioactive elements which, though in very small concentrations, are widely distributed over the earth's surface. The amount of radiation received by people from these sources is, however, highly variable because of the uneven distribution of radioactive substances in different rocks and different soils. Thus, the radioactive elements radium, thorium, and a radioactive isotope of potassium (K-40) occur in greater concentration in granite than in basalt, and in the latter in greater concentration than in sedimentary rocks. External radiation rates are lower in buildings constructed of wood than in those made of brick, and in the latter lower than in those of concrete or granite. Because of these variations the total radiation doses received from sources external to the body are rather different in different places.

Background radiation levels appreciably higher than the crude average of about 2,500 mrad per 30 years (one human generation) to which many populations are exposed occur, however, in some parts of the world. This is particularly true of a region in the state of Kerala, India, and in the state of Espírito Santo, Brazil, where many people inhabit regions of so-called monazitic sands, which contain the radioactive mineral thorium. The population exposed to these high radiation levels in Kerala is estimated to number about 100,000 persons. The average radiation level in Espírito Santo is around 500 mrad per year (15,000 per 30 years), while in Kerala estimates between 131 and as high as 2,814 mrad per year (about 4,000 to 84,000 mrad per 30 years) have been obtained.

Minute quantities of radioactive elements occur normally as constituents of human bodies and, of course, of all other living creatures. A certain number of atomic disintegrations occurs, therefore, within our bodies, and certain amounts of high-energy radiations are produced by these internal sources. Thus, the body of a man weighing about 150 pounds contains approximately 17 milligrams of radioactive potassium (K-40), and it has been estimated that human gonads receive some 500 mrad of gamma radiation from this source in 30 years. The presence of even smaller amounts of a radioactive isotope of carbon (C-14) exposes human gonads to about 1.6 mrad per year (about 50 per 30 years). The Report of the U.N. Scientific Committee on the Effects of Atomic

Radiation estimates the average total dose of radiation received by human gonads from both external and internal radiation sources as 100 mrem per year, or 3,000 mrem or 3 rem per 30 years.

Man-Made Radiations

Three rem per generation may be regarded as close to the irreducible minimum of background radiation to which mankind was, is, and will continue to be exposed. In addition to this rock-bottom minimal exposure, mankind is, and probably will continue to be exposed to some radiations from man-made sources. Among the latter, the radioactive fallout from the testing of nuclear weapons has in past years been discussed in learned articles, popular books, newspapers, and magazines, and from political tribunes. It may, therefore, surprise many readers to learn that the amount of exposure from fallout products is decidedly smaller than that from less publicized man-made sources — namely, from sources used for medical and industrial programs.

The story of the fallout is, very briefly, as follows. The explosion of atomic bombs and superbombs generates numerous unstable and radioactive isotopes of many chemical elements. Among these, strontium-90 and cesium-137 are considered most important, both because these radioactive atoms are produced in large amounts and because they decompose rather slowly, continuing to give off dangerous high-energy radiations for a long time. Strontium-90 has a half-life of 28 years; cesium-137 a half-life of 27 years. The half-life of a radioactive element is the length of time required for one half of the substance to decompose and to lose its radioactivity. But after two half-lives — that is, after 56 years, one quarter of the original radioactivity of the strontium-90 fallout will be preserved; after 84 years one eighth, etc. Last but not least, strontium-90 and cesium-137 are important because they are absorbed and retained in the bodies of plants, animals, and men.

Atomic explosions hurl radioactive materials high into the atmosphere, even into the stratosphere. These materials are carried aloft and then, for a certain time, transported about the earth by winds. Eventually they fall once more to the surface of the earth — hence, the term fallout. A part of fallout is "local" fallout; these are materials which settle in the neighborhood of the test site within a few days after the explosion. But lighter particles in the upper reaches of the atmosphere and in the stratosphere may remain aloft for several years, may be carried around the whole earth, and may settle only slowly — but not slowly enough to have lost their radioactivity. Since most of the atomic test sites have thus far been located in the temperate and subtropical belts of the Northern Hemisphere, fallout in this region has been greater than in the Arctic, along the equator, or in the Southern Hemisphere. Nevertheless, fallout is a world-wide problem.

Some strontium-90 and cesium-137 may enter the human body through actual inhalation of air containing fallout particles. By far the most important source, however, is vegetables that have absorbed fallout products from the soil, or milk and milk products from cows that have been pastured on contaminated vegetation. A great deal of research in several countries has attempted to estimate the amounts of strontium-90 and cesium-137 which people accumulate, and the amounts of ionizing radiations which will reach human sex cells from these radioactive atoms. Indeed, strontium-90 has made its unwelcome appearance in milk, in cereals, and in vegetables; and it can be detected in the bones of persons of various ages, the concentration being especially high in the bones of growing children.

The radiation exposure of human sex cells through the testing of atomic weapons amounts to only a fraction of the presumably irreducible exposure to radiation from natural sources. This does not mean that the additional radiation is negligible. Moreover, if a war were unleashed in which nuclear weapons were used, the radiation exposure would be much greater than that resulting from either the testing of weapons or natural (background) sources.

By far the most important source of radiation to which human populations are at present exposed is neither background radiation nor atomic bombs but the x-ray machines in hospitals and in doctors' offices. X-rays have a very important place in modern medical practice. They are virtually indispensable in diagnosing many ailments, from toothache to tuberculoisis and cancer. They are essential preliminaries for many surgical operations. Equally important is the therapeutic use of radiation. Treatments with radiation are applied in cases of malignant tumors and for many other lesser ailments. Radioactive substances, particularly iodine-131, phosphorus-32, and gold-198, are given internally to some patients, usually in small quantities, and chiefly for diagnostic purposes.

The unquestionably beneficial uses of radiation in medical practice are, unfortunately, coupled with some attendant danger of genetic damage. Genetic damage arises when high-energy radiations impinge on reproductive organs. This occurs to some extent no matter what part of the body is irradiated intentionally. If a beam of sunlight enters a room, part of the light is reflected and scattered throughout the room. Similar scattering occurs also with x-rays and other radiations. When, for example, a beam of x-rays is directed at the chest of a patient to examine his lungs, a fraction of the rays also reaches the sex cells. This occurs even with dental x-rays. When x-rays are directed at the pelvic region, as in obstetrical examinations, the gonads of the mother as well as those of the unborn child may be directly in the intense field of the primary beam.

More and more persons are employed by industrial plants that use x or gamma radiations to inspect their products. One reads with increasing frequency of this or that company which has obtained a cobalt-60 source of radiation for some aspect of industrial research. Quality control through automatic machinery activated by radiation sources is becoming commonplace. Finally, both in industrial and research laboratories the use of radioactive atoms has become extremely common. Radioactive atoms do not differ from ordinary atoms in their chemical behavior. Consequently, in the study of industrial or physiological chemical processes, radioactivity offers a means of tagging certain atoms and following them through a variety of reactions. The technique used is basically the same whether one wishes to determine what compound in gasoline deposits carbon on valves in automobile engines or what part of a parasitic virus enters and kills a bacterium. For the genetic effect of radiation on human populations, it makes no difference what problem is being investigated; the increased use of radioactive isotopes means that more members of the population are being exposed to radiation.

How much radiation exposure is incurred by people because of the medical, occupational, and technological use of radioactive substances? This is very difficult to estimate with any precision. The sources of uncertainty are many. First of all, it is evident that people in technologically advanced countries are exposed to radiations more frequently than people in primitive or underdeveloped countries. But even in countries with well-organized research facilities it is no easy matter to find out the genetically important exposure dose — that is, the dose which reaches the sex cells. Take, for example, the radiations used to treat patients with malignant growths. Even if massive doses of radiation are applied, the genetic effect may be slight; indeed, most such patients are beyond the childbearing age, and for many of them even the chances of survival are small.

The Report of the U. N. Scientific Committee on the Effects of Atomic Radiation has collated and attempted to interpret a mass of data on man-made radiation exposures in Austria, Denmark, England, France, Japan, Sweden, and the United States. The figures arrived at vary from 20 to 150 mrem of exposure per year from diagnostic radiations; from 1 to 30 mrem from therapeutic exposure; less than 1 mrem per year from radioactive substances taken internally; and less than 2 mrem per year from occupational exposure. These are averages per person; certain individuals of course receive considerably more than others. Multiplying these figures by 30 to obtain genetically effective doses per human generation, we obtain totals of 720 mrem to 5,500 mrem. In other words, people in technologically advanced countries are, on the average, exposed to a total amount of man-made radiation about twice as great as the average natural background.

This trebling of the natural radiation exposure is certainly disquieting, since it is bound to produce some genetic damage. Most certainly, we should not be frightened to the point of denying people the benefits which legitimate medical uses of radiation are capable of giving. These benefits are too obvious and too important to be renounced. But it is also certain that precautions should be taken to minimize radiation exposure, particularly of the sex cells, as far as possible. This can be done by using screens made of materials opaque to x-rays to shield the reproductive organs of patients as well as physicians and medical technicians. Recognition of the genetic and other dangers of radiation should make us avoid unnecessary exposure. It is, for example, very doubtful whether x-ray machines should be allowed to be used for shoe-fitting. One need not rely solely on genetic arguments to justify this statement; the growing bones in children's feet, because of the rapid division of bone cells, are especially sensitive to x-ray exposure, just as are other tissues — including cancerous growths — in which cell divisions are common. Solely on genetic grounds one can argue, however, that temporary male sterility should never be induced by irradiation; too many alternative and convenient contraceptive techniques are available to justify the use of an agent which may doom a future individual to a life of misery and unhappiness.

Radiation in
Research and Industry

According to unsubstantiated folklore, the first experimental use of an atomic tracer occurred in a Hungarian boarding house where Dr. Georg von Hevesy was a student roomer. He suspected that the Thursday-night stew was concocted from Monday-night's leftover meat, although his landlady had self-righteously denied that his suspicions had any basis in fact. Hevesy at the time was working with "heavy hydrogen," deuterium. Deuterium is a rare substance, and so Hevesy conceived the idea of pouring a small quantity of it into Monday's meat dish and then testing a sample of the following Thursday's stew for its presence. The experiment worked; Thursday's stew contained the tracer, and the landlady was confronted with indisputable proof of her duplicity. Folklore does not tell us whether the quality of Thursday's evening meal improved; it merely records that the first use of a tracer was in the successful demolition of one old wife's tale.

Radioactive isotopes, in a far more convenient manner than deuterium, also serve as atomic tracers. The radiation emitted by these unstable atoms is the very heart of their convenience; the presence of these atoms can be revealed by radiation detection devices; of which the simplest is photographic film. Now, we admit that radiation is hazardous to life and health. What, then, are some of the applications to which radiation and radioactive isotopes are put that justify the hazards their very use entails? In this essay we shall consider applications frequently encountered in science and industry. Not discussed in this essay are the medical uses of radiation; in a general way at least, all of us are aware of the diagnostic X-ray pictures used by doctors and dentists, of fluoroscopic examinations, and of cancer therapy.

On several occasions the words "food chain" and "food web" have been used to describe the flow of energy from its capture by green plants and storage in starch, through the herbivores that graze on green plants, through carnivores — large and small — that eat herbivores or other carnivores, to the decomposers who once again release water and carbon dioxide to the environment. It is, of course, water and carbon dioxide that are combined at the cost of much energy (supplied by sunlight) by the green plant to make starch; it is this energy that is released in small packets at every step throughout the entire food web until the last bit is released in the final production of carbon dioxide and water once more.

The above description of energy flow is a description done in bold strokes indeed. A multitide of details has been ignored. Green plants are eaten by herbivores, but how is plant "stuff" turned into animal "stuff?" Why is food chewed so thoroughly? What happens to it in the stomach? In the intestine? What are the functions of the many digestive glands? The chemical events that are involved in the digestion of a cheeseburger and the utilization of this food for energy and growth by a teen-ager lie beyond the total abilities of the most complex chemical industry imaginable. Nevertheless, biochemists have unraveled many of these events step by step until our knowledge of them are amazingly complete. Radioactive isotopes have been an indispensible tool in these painstaking analyses.

Photosynthesis, the production of sugar from carbon dioxide and water through the utilization of light energy, can serve as an example of a chemical process whose analysis would have been impossible without the radioactive carbon isotope, carbon-fourteen (C-14; the chemical symbol for "normal" carbon is C-12). The photosynthetic process was presented earlier in this book as follows:

$$\text{Carbon dioxide} + \text{Water} + \text{Light energy} \longrightarrow \text{Starch} + \text{Oxygen}$$

This description is true only in respect to the bare essentials: the input, the energy requirement, and the outcome. The actual chemical events, however, are not to be reconstructed from this simple description; they are more accurately described in the following words:

> The green plant, through its own metabolic machinery, provides a sugar that contains five carbon atoms (a "five-carbon" sugar, in short). This sugar combines with carbon dioxide to form an unstable six-carbon sugar that decomposes almost immediately into two three-carbon molecules. The latter molecules are the ones that combine eventually to form the sugar, $C_6H_{12}O_6$. This reaction requires both the energy the plant has captured from sunlight and a hydrogen atom from water. It is the utilization of the hydrogen from water that makes oxygen available for eventual release into the atmosphere.

The analysis of the complex pathway has been made possible by the simple device of using radioactive carbon in the carbon dioxide supplied to the plant. By tagging this particular carbon atom, it is then possible to follow it wherever it goes as it joins the five-carbon sugar, or as it becomes diluted (in relative terms) by the splitting of the six-carbon compound into two identical three-carbon ones. In the latter case, only half of the three-carbon molecules are labeled since each six-carbon molecule had but one radioactive carbon atom. The radioactivity of atoms introduced experimentally from one source allows the chemist to recognize these very same atoms in all subsequent reactions. Furthermore, by mounting thin slices of material on fine photographic film, he can also learn where within the plant cells these complex reactions occur because the disintegrating atoms develop the nearby photographic emulsion; this is photography on a fine scale indeed.

Molecular biology furnishes the second example of the use of radioactive atoms in research. The ultimate control of protein synthesis is known to reside in DNA; that is, the linear sequence of purine and pyrimidine (also known as "nitrogenous") bases in DNA molecules is transcribed to a corresponding sequence in messenger-RNA. The messenger-RNA sequence is then translated by an appropriate mechanism into a sequence of amino acids – the backbone of a protein molecule. The eventual three-dimensional configuration of a protein molecule is determined – or, at least, one of several such stable configurations is determined – by the actual sequence of amino acids. There may be more than one stable, three-dimensional configuration for a given sequence of amino acids just as there is more than one stable resting position for an ordinary brick.

Despite the general knowledge about the relationship of DNA structure and amino acid sequences in proteins, the information available does not tell us at which end a protein molecule is started. Is there a given end at which to start reading the instructions contained within the messenger-RNA? Or can these instructions be read from either end in either direction? If we consider simple household recipes, we can see that some steps call for combinations of ingredients that can be assembled in any order and then mixed, but others, such as those for the icing of a cake, presuppose a certain order in which things are accomplished; that is, the cake is baked first.

The use of radioactive isotopes allows one to determine the direction in which protein molecules are synthesized. Suppose that it is known that a protein such as hemoglobin can be written $A \cdot B \cdot C \cdot D \cdot E \cdot F \cdot G \cdot H \cdot \ldots X \cdot Y \cdot Z$ where all 26 letters of the alphabet stand for successive amino acids. What is not known is whether the molecule is put together $A \cdot B \cdot C \cdot D \ldots$, or $Z \cdot Y \cdot X \cdot W \cdot V \ldots$, or both ways. The decision concerning the direction of protein systhesis can be made if labeled amino acids (amino acids containing carbon-14) are provided for a system of cells that are actively making protein. Such a system is the one found in cells that make hemoglobin; these cells do very little else. If such cells are exposed to an artificial growth medium in which all available amino acids carry carbon-14, there are two major possibilities for the distribution of radioactive carbon in the finished hemoglobin molecules. If the protein molecule is put together from the A-end by adding B, C, D, and so forth until X, Y, and Z are attached, then the switch from normal amino acids to those with radioactive carbon atoms will cause the addition of radioactive Z in some, nearly finished molecules; Y and Z in others, X, Y, and Z in still others. When a collection of hemoglobin molecules is analyzed, the amount of radioactivity will be greatest in what were unfinished ends of molecules at the start of the experiment. The amount of radioactivity will decrease as one moves toward the end of the molecule at which synthesis starts.

The second possibility is the same except for the reversal of letters. If the molecule is made from Z toward A, molecules that were nearly finished at the introduction of amino acids containing radioactive carbon atoms will incorporate these labeled amino acids at the A end. And so, some molecules will have A labeled by radioactivity, others will have B and A, still others C, B, and A, and so forth. Once more the direction in which the molecule is assembled can be inferred by the decreasing amounts of radioactivity detected in the segments A, B, C, D, and so forth.

A third possibility is, of course, that protein synthesis can begin at any point within the molecule and proceed in either direction. This would lead to the incorporation of carbon-14-labeled amino acids at any point along the molecule and would lead to a uniform distribution of carbon-14 atoms along the length of the hemoglobin molecule.

These are the sorts of problems that scientists are able to solve with the use of radioactive atoms. They are fundamental problems; their solution gives us deeper insight into ourselves and our surroundings. The callous and unthinking can, of course, reword the same problems in the most ridiculous ways. A recent Secretary of Defense scornfully disclaimed any interest in "Why grass is green." As we have seen, however, the greenness of grass is responsible for all of the atmospheric oxygen; for all the coal, oil, and gas on earth; and makes possible all known forms of life. The greenness of grass is not to be scorned!

Industry also has problems to which radiation and radiation-emitters offer solutions. Quality control lends itself to radiation control. The thickness of any material that is extruded between rollers can be measured by the penetration of appropriately chosen atomic radiation; the amount of radiation that penetrates is measured by an appropriate device and is used to effect an immediate adjustment of the pressure exerted by the extruding apparatus. Radiation through its detection devices plays the same role in the operation of these machines as does heat when it affects a thermostat and in this way controls the furnace.

There are any number of industrial uses for radioactive tracers. Which, for example, of many carbon-containing molecules of gasoline are responsible for the carbon deposit in an automobile engine? All? Or just a few? Or one? The answer can be obtained by labeling one compound after another with carbon-14 and then, following an experimental run of the engine, testing the carbon deposit in the engine for radioactivity.

X rays are used during manufacturing processes much as a doctor uses them for the diagnostic examination of patients; flaws in a variety of manufactured products are revealed by X-ray examination. This type of examination is especially useful under circumstances where *every* item (rather than a representative sample of items) must be examined before use. X-ray examinations are used, too, for large machines that would be tremendously expensive to dismantle for a close visual examination of each part.

Although the questions to which industry addresses itself are not necessarily fundamental ones in the sense of "pure" research, they are nevertheless valuable ones. On the other hand, many "fundamental" problems of pure research are trivial. These points are made to emphasize once more that radiation and radioactive isotopes are hazardous to health and are to be used only under strict control, only in attempting to solve worthwhile questions, and only if other techniques are inadequate. A recent item in the *New York Times* reported on the development of a pocket-size X-ray device. The item concludes by saying, "According to the company, the Soremark X-ray unit has been tried for dental X rays and X rays of other parts of the body without harmful effect." A pocket-size X-ray emitter seems to me to be one of the machines the world can well do without. The world will do *with* them in great numbers, however, if the harm X rays can cause is not widely known.

The Immediate Effects
of Radiation Exposure

Radiation is extremely effective in damaging living cells; the energy required to kill a man by radiation is much less than that (in the form of heat) required to warm a cup of tea. This brief essay will dwell on the empirical observations related to radiation exposure and will offer partial explanations for the observed symptoms in terms of the molecular effects of ionizing radiation. The account presented here cannot be complete; the technical aspects of radiation biology are too complex to be adequately summarized in simple form.

Dr. Roentgen reported his discovery of X rays ("X" for "unknown") in late 1895; within four months a report on the loss of hair through the exposure of a person to X rays had been published. The list of effects caused by exposure to X rays grew rapidly thereafter: reddening of exposed skin (once used to measure radiation exposure!), ulceration, the appearance of tumors, and skin cancer. Experimentalists found that X rays killed bacteria and, for many different organisms, either killed developing eggs or led to grossly deformed individuals among the successful survivors.

Living systems are aqueous systems; 70 per cent or more of cellular material is water. The immediate effect of radiation appears to be the production of highly reactive radicals from molecules of water, H_2O. These new types of molecules — OH and H, for example — are exceedingly reactive. In pure water these radicals would rejoin to form H_2O once more or would misjoin to form H_2 (hydrogen gas) and H_2O_2, which would then break down into water and O_2 (molecular oxygen). Thus, the irradiation of water is known to generate the gases, hydrogen and oxygen.

Within exposed cells the OH and H radicals can react with normal cellular materials. They can react with proteins, for example, or with the genetic material, DNA. These reactions represent a double-headed attack on the cell's existence; not only are proteins destroyed but the machinery by which undamaged proteins might be made anew is destroyed as well.

Protein is literally the stuff of life. Virtually all reactions that provide life-sustaining energy are mediated by enzymes, protein catalysts. Many of these reactions take place at definite sites within cells, sites specified by proteinaceous membranes. Membranes also determine what sorts of chemicals can pass from one part of the cell to another; sodium may be allowed to pass through one membrane whereas potassium, a very similar atom, is prevented from doing so.

The role a protein plays in the life of a cell is determined by its three-dimensional shape, and by the alterations this shape undergoes as one step after another occurs within the sequence of events the protein controls. The attachment of novel, unexpectedly reactive molecules to proteins and the accompanying alteration of protein structure lead to the inactivation of enzymes and to the destruction of membranes. If the damage is severe enough, death of the cell quickly ensues.

The genetic material, DNA, is also subject to destruction by irradiation. DNA molecules can be broken, can rejoin in aberrant ways, can lose segments, or can be converted into organic peroxides by the attachment of reactive molecules. This damage can prevent the cell from making new protein molecules to replace damaged or missing ones because, as we have seen elsewhere, DNA contains the information on how to build these large molecules. Radiation damage to DNA renders its information useless. Damage of this sort is comparable to the destruction of a library. Even if the cell manages to survive the immediate destruction of protein molecules, it may not be able to divide properly because of physical damage to its chromosomes. Broken chromosomes may rejoin in such a manner that they are unable to furnish each daughter cell with a full complement of DNA. In this case, the daughter cell will die. This type of damage is especially severe in cells that are actively dividing such as those cells that are responsible for replacing others that are lost or destroyed in carrying out their normal functions.

Many of the symptoms developed by Mrs. Nakamura and her children during the month following the atom-bombing of Hiroshima can be understood in terms of radiation damage to various tissues; others are even now poorly understood. The thirst developed by the entire family almost immediately after arrival at Asano Park was probably caused by a loss of water into the intestine; diarrhea is a common affliction following radiation injury of the stomach and intestine. Experimental work with rats has shown that the movement of food in the digestive tract is upset by radiation; the movement may be reversed, an effect that in man leads to vomiting and nausea.

Tissues in which cells are dividing regularly and rapidly are among those most sensitive to radiation exposure. Tissues of this sort include the lining of the gut because old cells are constantly being sloughed off into the lumen of the gut, only to be replaced by new ones. And in bone marrow, the blood-forming cells are always forming new red blood cells to replace those that are lost from circulation through wear and tear. (The average red blood cell lasts only 120 days in circulation; since the total number of red blood cells in an adult is about 30 trillion, approximately 300 billion new red blood cells must be made each day to replace that day's loss.) The cells in the individual hair follicles also divide continuously, and so these, too, are subject to extensive radiation damage.

The list of body tissues that are sensitive to radiation corresponds very nearly to the list of symptoms exhibited by an irradiated person: intestinal disturbances frequently accompanied by infection, weakness because of severe anemia, and loss of hair following the destruction of hair follicle cells. These symptoms are expected to set in approximately two or three weeks following exposure to radiation. In the case of the Nakamuras the symptoms should have occurred between August 20 and August 27 because the bomb was exploded on August 6. In the account given by John Hersey, Mrs. Nakamura's hair began falling out on August 20; on August 26 both Mrs. Nakamura and Myeko were extremely tired. The other children, Toshio and Yaeko, were not ill on the 26th although they did lose some hair later. In rechecking the events as they befell the Nakamura family on that fateful August 6, I see that Toshio and Yaeko were buried in the rubble of their home more deeply than their mother and younger sister: this temporary burial of the two older children may have shielded them from some of the radiation to which their mother and young sister were exposed.

Radiation
and the Unborn Child

Cells most sensitive to radiation damage are those that are actively engaged in cell division; these cells are subject to damage because of the misdivision of damaged chromosomes during cell division and an inability of the cell's metabolic machinery to make or replace the many proteins that are needed in preparing for subsequent cell divisions. In the case of adult persons, extremely sensitive tissues — blood-forming tissue, the lining of the gut, hair follicles, and others — can be enumerated quite simply because there are so few of them. For young children the list of sensitive tissues would be extended to include the growing areas of various long bones such as the arm, leg, finger, and toe bones; it is quite possible, for example, that the three Nakamura children are smaller adults today than they would have been had they not lived in Hiroshima during August, 1945. The youngest child, Myeko, may easily have suffered from retarded growth in later years because of radiation damage to her thyroid gland.

If radiation is harmful to adults and growing children, however, it is disastrous to the unborn child because a young embryo consists of little else than rapidly dividing cells. For this reason, the developing embryo represents a system quite different from that of either adults or children and deserves an essay of its own.

The calendar of events for the developing human embryo proceeds about as follows:

Day 0   An egg (weight: two-millionths of a gram) is fertilized by a sperm (weight: negligible).

Day 10   The embryo, now a mass of cells resembling a small raspberry, becomes embedded in the wall of the uterus.

Day 40   Major organ systems are now complete and the embryo, at least superficially, resembles that of nearly any other higher animal from fish to mammal.

Day 60   Face complete; fingers and toes separated; tail diminishing. Total weight: 2 grams (about 1 million times the weight of the original egg).

Day 60-270   In the period between day 60 and birth, the embryo grows nearly 2000-fold in size; at birth the baby weighs about 3400 grams or 7½ pounds.

An embryo does not grow uniformly; the head and brain are the first parts to form. Development spreads posteriorly; the forelimbs (arms) take form, for example, before the hindlimbs (legs). Areas of differentiation by their very nature are areas of rapid cell division. Consequently, if radiation strikes a developing embryo at a given moment, the most severely damaged portion is that which is differentiating at that instant.

During the first two weeks after fertilization, an embryo is merely a mass of dividing cells with little or no organization in the sense of recognizable adult systems. Irradiation of the embryo at this time leads to its death or, should development continue, to extremely severe physical abnormalities.

Between the time of the embryo's implantation in the uterine wall and the 40th day when most organ systems are complete, even low levels of irradiation can cause severe harm to one part or another of the developing child; the injured part is in each instance the one that was undergoing differentiation at the moment of radiation exposure. This effect is so precise that the developmental embryology of an experimental animal can be studied by irradiating pregnant females at various times and noting the resulting abnormalities in their offspring.

Many embryos carried by pregnant women of Hiroshima and Nagasaki (the site of the second atomic bomb explosion) were killed by radiation exposure. Of the surviving births, from one-fourth to one-half were mentally retarded. Mrs. Nakamura was not pregnant at the time Hiroshima was destroyed; had she been, the embryo would most likely have suffered severe harm because of the high level of radiation to which she was exposed. The sensitivity of growing embryos to harm through radiation exposure is such that a number of governments now recommend a therapeutic abortion if the mother's abdomen has been exposed during pregnancy to as little as 10 roentgen of penetrating radiation.

Radiation and Cancer

Death by cancer has been a common fate of persons who work with radiation or radioactive material. Both Marie Curie, who, with her husband, discovered radium, and her daughter Irene died of illnesses induced by radiation exposure. The same was true of many pioneer scientists who worked with crude X-ray machines using primitive techniques. It was also true of many early dentists, especially those who insisted on holding their patients' dental films themselves, for these dentists were exposed to X rays time and again each day. It is true even today, however, of nuclear physicists who should have access to the most up-to-date mechanical and safety equipment; a number of persons whose careers included pioneer work leading to the development of nuclear reactors have died of leukemia.

The survivors of the atom bomb explosions of Hiroshima and Nagasaki have succumbed to leukemia at an annual rate ten or fifteen times higher than the national average of Japan.

In 1954 a thermonuclear device (an H-bomb) was tested at Bikini in the South Pacific. The blast, the equivalent of 15 million tons of TNT, was larger than expected; the mushroom cloud climbed higher than expected; and fallout debris was unexpectedly carried eastward by high level winds rather than westward. Four of the Marshall Islands that were inhabited were subjected to severe radioactive fallout. The Marshallese were evacuated to a nearby atoll where they remained for three years before being allowed to return home. As a consequence of their exposure, these persons have been subjected to frequent, thorough medical examinations.

The number of Marshallese exposed to radiation was too small to yield satisfactory information on those consequences of radiation exposure that are expected to affect only a tiny fraction of exposed persons. All of the exposed natives suffered from superficial radiation burns. Two or three of them died of cancer soon after exposure but as far as one knows these deaths were not related to the radiation exposure. There is little uncertainty, however, concerning the children who were five years or less at the time of the 1954 accident. Many of these young persons are retarded in growth; many have developed tumorous nodules on their thyroid glands. It appears that thyroids of children less than ten years of age are exceptionally sensitive to radiation and that exposure of the young Marshallese is now resulting, in almost all exposed children, in thyroid tumors.

Radiologists, patients undergoing radiation treatment, and children irradiated before birth while still in their mothers' wombs die more frequently of leukemia than do other members of the population. The death rate from leukemia of old-time radiologists is about three times the national average. Persons suffering from certain rheumatic diseases for which heavy therapeutic radiation is prescribed have a death rate from leukemia nearly ten times the national average; in this case, however, the rheumatic disease itself might be related to the leukemia that eventually develops.

The manner in which radiation induces cancerous growths is virtually unknown. Geneticists at various times have suggested that a cancer cell is a mutant cell, that cancer represents a somatic mutation (a mutation in one of the cells of which the body is composed). Cancer specialists, for a number of cogent reasons, have opposed this view. In recent years virus particles have been identified with an increasing number of types of cancerous cells. It is an experimental fact that viruses that have been incorporated into the DNA of their host cells can be induced to separate and undergo autonomous growth by radiation exposure; as in the case of the induction of gene mutations, the frequency with which virus particles are "activated" within their carrier cells by radiation increases linearly with radiation exposure.*

A person suffering from cancer worries very little about the induction of new cancerous cells when his doctor recommends radiation therapy. How can radiation play the role of both villain and hero in respect to cancer? How can cancers be both caused and cured by the same radiation?

The use of radiation for the cure of cancer depends upon the biological characteristics of cancer cells. These are rapidly dividing cells and, hence, are subject to radiation-induced chromosomal damage that we mentioned earlier. The treatment is especially effective if the cancer is localized. If, however, the cancer has sloughed off malignant cells (cells that travel throughout the body and initiate scattered cancerous growths at widely separated spots), radiation treatment cannot effect a cure. Radiation may still serve as a deterrent for an advanced tumor, however, despite the possibility that the treatment may miss numerous smaller secondary malignancies. On the other hand, tumorous cells that have a special affinity for certain elements or compounds can sometimes be tricked into taking up radioactive isotopes and, in a sense, committing suicide. Radioactive iodine is effective, for example, in treating thyroid tumors.

This essay may be terminated with two messages. In the proper hands, radiation can arrest either momentarily or permanently an otherwise incurable cancer. In careless hands — whether scientific, industrial, or military — radiation can cause leukemia and other forms of cancer. Radiation is deadly! Elaborate safety regulations have been set up by both federal and state governments to guard against the misuse of radiation and radioactive materials; be sure that these are obeyed in your neighborhood!

*See also the essay entitled "The Biology of Cancer" in Vol. III.

Radiation
and Future Generations

Whether we admit it or not and whether we abide by it or not, the standard of behavior that makes civilizations possible is that which Christians know as the Golden Rule: "Do unto others as you would have others do unto you." Wherever civilizations have flourished and wherever persons have been thrown into intimate contact with one another, a rule of this sort has evolved as a code of behavior. In India the comparable rule is stated: "Do naught to others which if done to thee would cause thee pain." In China the advice is: "Not to do unto others as you would not wish done unto yourself."

Rules of this sort are essential in urban societies. In effect they exhort each person to look upon all persons, including himself, as equals. These rules discard the argument that *this* act is different from that act because this one was done to *me* while that one was done to him. Urban life is possible only because of the Golden Rule (or its equivalent) and forgiveness. In legal parlance "forgiveness" becomes "statute of limitations"; in a sparsely inhabited area "forgiveness" is replaced by "forgetfulness." The Golden Rule minimizes conflicts; forgiveness dampens the action-reaction cycle in case a conflict between persons does arise.

The notion that all persons are equal is not restricted to persons living simultaneously; it applies to human beings, born or unborn. We cannot in good conscience do today whatever might give us momentary pleasure or profit and say, "To hell with the next generation." We cannot do this because we today look at some of the ills that we have inherited and because of them we damn our predecessors. "Do unto others" refers to coming generations just as much as it does to our own.

We pass on to our descendants a world in which to live and DNA with which to make something of themselves – literally. Because of the numbers of persons now living and the technological demands of industrial societies, the world is becoming an ecologically simpler and simpler (and, hence, more unstable) place; natural diversity of the environment is being replaced by a man-made uniformity. We are playing, therefore, with an unstable world; whether or not we pass it on to our remote descendants depends upon whether or not we accidentally drop it.

The DNA we pass on to coming generations is, ostensibly, that which we received from our ancestors. The DNA we pass on, in fact, undergoes mutational changes while it is in our possession. Some of the mutations are "spontaneous" in origin; that is, these changes occur as the result of thermal instability or because of inherent instabilities of genetic material. Added to spontaneous mutations are those that are induced by ionizing radiation. Again, some of this radiation is natural; it represents radiation to which mankind has always been exposed. In effect, these induced mutations can be added to the store of spontaneous ones that occur during each of our life times.

In addition to all other gene mutations, however, are those that are induced by man-made radiation. These are the mutations induced in the germ cells of many different persons by dentists during dental examinations, by mobile X-ray units searching for tuberculosis, by X rays used for the inspection of industrial products, and by the military use of atomic and hydrogen bombs. These mutations are passed on to today's children and to tomorrow's children's children. What are the effects of these mutations? How frequent are they? What is their eventual fate? These are the sorts of questions to be considered in this essay.

A gene mutation is a change in DNA that destroys the normal function of a gene; a mutant gene may produce an abnormal protein molecule, no protein molecule at all, or the right molecule but at the wrong time. Such changes may be caused by the substitution of one base pair for another in the DNA molecule, by the deletion of one or more base pairs, or even by the insertion of new pairs. Alterations of these sorts in the DNA molecule do not interfere with the mechanics of gene replication; instead, they alter the genetic message that specifies the sequence of amino acids in the corresponding protein molecule.

The absence of an enzyme or the presence of an enzymatically inactive protein molecule can have dire consequences for the affected individual. Some children are born, for example, lacking the enzyme that is necessary for the conversion of phenylalanine to tyrosine, that is, for the conversion of one particular amino acid to another. Because phenylalanine cannot be converted in these individuals, it accumulates within their bodies and is excreted in large amounts (together with another of its breakdown products, phenylpyruvic acid) in their urine. The disease caused by this defect is known as *phenylketonuria*; among other characteristics, afflicted persons suffer from extreme mental retardation.

Mutations arise only rarely at any one gene locus. Stated somewhat differently, the ability of DNA to replicate is so precise that a segment containing 300 base pairs (enough to carry instructions for linking together 100 amino acid molecules in the unique order needed to make a particular protein molecule) will be copied erroneously only once in about 100,000 times. There are, however, many genes of this sort within the set carried by a single individual and so the chance that at least one gene in the entire set has a newly arisen error is quite large — one chance in ten perhaps.

Mutant genes are lost from a population through the failure of their carriers to reproduce. A person suffering from phenylketonuria, for example, leaves no offspring; consequently, the mutant genes responsible for phenylketonuria are not passed onto the next generation through an affected individual's line of descent. Without attempting to prove the claim mathematically, I shall say only that mutant genes accumulate in populations until their loss by virtue of the reproductive failings of their carriers equals the rate at which they arise in the population by mutation. The rates of origin and elimination must be equal if the frequency of the mutant gene in the population is to be stable; the frequency of affected individuals in the population increases to whatever level is necessary to bring about the required rate of elimination. By analogy, if a faucet drips into a bathtub whose plug is improperly set, water will accumulate in the tub until water escapes around the plug at the same rate at which it drips from the faucet. The depth of water in the tub is responsible for setting the rate at which water leaks past the faulty plug; the deeper the water, the faster it escapes down the drain.

Radiation affects future generations because it induces gene mutations. A relatively low average exposure given to the entire human population would double the present spontaneous mutation rate. This exposure (sometimes called the "doubling dose") has been estimated to be as low as 10 roentgen; it might be given over an interval of one generation (30 years) at the rate of one-third roentgen per year. If a population were to be exposed continuously to a doubling dose of irradiation, the frequency of carriers of mutant genes in the population would increase until the rate of elimination of mutant genes once more equalled their (now higher) rate of origin. Phenylketonuria which now affects about 100 of every million newborn children might increase until 200 of every million were affected. At least as an approximation, the frequency of persons affected by other mutant genes with severe effects on their carriers would also be doubled.

At times the discussion of the mutagenic action of high-energy radiation has been presented as if the continued existence of mankind were at stake. This is not really the issue. Mankind will survive mutagenic radiation; or if he does not survive, mutant genes alone will not carry him off – blast, heat, and total body irradiation might. The important point I want to emphasize here is that exposure of persons to radiation can be – and should be – minimized by exercising scrupulous care. Fast film can be used for diagnostic X rays so that exposure times can be lowered. Protective aprons can be worn over the gonads so that stray radiation will not affect the germ cells of either patient or doctor. TV sets and other household devices that emit radiation can be built so that the radiation they generate is absorbed by safety screens and other devices. Trivial uses of X rays, such as their former inexcusable use in shoe stores and beauty parlors, can be abolished. These are acts of simple courtesy. They are sensible acts from the viewpoint of both public health and industrial safety. They minimize the chance of inducing cancers in today's population, and they reduce the odds that a child born several generations from now will be a genetic monster.

The use of radiation, like the use of natural resources, requires that the Golden Rule be applied not only to our relations with our contemporaries of today but also in the anticipation of our descendants of tomorrow.

Appendix A

The following essays in volumes I and III discuss matters related to the sections of this volume:

Genetics

Evolution

Appendix B

Table of Contents: Volume III

The Nervous System and its Organization: An Essay in Five Parts

1. The gross structure of the brain

2. The circuitry of the central nervous system

3. Memory: The storage and recall of information

4. The primitive brain

5. The individual nerve cell: To fire or not to fire

On the Inheritance of Behavior

On Kidding Ourselves

On Escaping Reality

On Obedience and Conformity

Epilogue

Index

An alphabetical listing (based on **KEY WORDS** of their titles) of the essays of this Volume

TECHNOLOGY and the future of man, 108

Radiation and the **UNBORN CHILD**, 204

On the biology of race: Genetic **VARIATION** between local populations, 150
On the biology of race: Genetic **VARIATION** within local populations, 150